The Death of Motoring?

The Death of Motoring?

CAR MAKING AND AUTOMOBILITY IN THE 21ST CENTURY

PAUL NIEUWENHUIS AND **PETER WELLS**
Cardiff Business School, University of Wales, Cardiff

JOHN WILEY & SONS
Chichester · New York · Weinheim · Brisbane · Singapore · Toronto

Copyright © 1997 by   John Wiley & Sons Ltd,
Baffins Lane, Chichester,
West Sussex PO19 1UD, England

National   01243 779777
International   (+44) 1243 779777

e-mail (for orders and customer service enquiries):
cs-books@wiley.co.uk
Visit our Home Page on http://www.wiley.co.uk
or http://www.wiley.com

Other Wiley Editorial Offices

John Wiley & Sons, Inc., 605 Third Avenue,
New York, NY 10158-0012, USA

WILEY-VCH Verlag GmbH, Pappelallee 3,
D-69469 Weinheim, Germany

Jacaranda Wiley Ltd, 33 Park Road, Milton,
Queensland 4064, Australia

John Wiley & Sons (Asia) Pte Ltd, 2 Clementi Loop #02-01,
Jin Xing Distripark, Singapore 129809

John Wiley & Sons (Canada) Ltd, 22 Worcester Road,
Rexdale, Ontario M9W 1L1, Canada

Library of Congress Cataloging-in-Publication Data

Nieuwenhuis, Paul.
  The death of motoring?: car making and automobility in the 21st
century / Paul Nieuwenhuis and Peter Wells.
    p. cm.
  Includes bibliographical references and index.
  ISBN 0-471-97084-0
  1. Automobiles–Design and construction. 2. Automobile industry
and trade. 3. Transportation, Automotive. I. Wells, Peter, Dr.
II. Title.
TL278.N54   1997   97-9815
629.2'31–dc21
                                                       CIP

British Library Cataloguing in Publication Data

A catalogue record for this book is available from the British Library

ISBN 0-471-97084-0

Typeset in 10/12pt Sabon from the authors' disks by MHL Typesetting Ltd, Coventry
Printed and bound in Great Britain by Biddles Ltd, Guildford and King's Lynn
This book is printed on acid-free paper responsibly manufactured from sustainable
forestation, for which at least two trees are planted for each one used for paper production.

Contents

1 Environmental and Economic Limits on the Automotive Industry

... For, the Visitor concluded,
If you were conceived in a car
As many are,
If you first made love in a car,
As many have,
If you went to work in a car,
As many do,
And if you derive your sense of freedom from cars ...
You are going to defend them
To the death
          (Heathcote Williams, 1991, *Autogeddon*: 81)

1.1 INTRODUCTION

This book draws upon the wide range of projects undertaken at the Centre for Automotive Industry Research in Cardiff, where a persistent theme has been the impact of environmental issues on the sector (Nieuwenhuis et al., 1992; Nieuwenhuis & Wells, 1994), to provide a richly textured account of the industry as it exists today and its likely shape in the future. In effect, we are presenting our vision of the industry for the future. Necessarily this vision is personal, perhaps rather eclectic and, given the size and complexity of the industry, partial. None the less, the book marks the first attempt to combine such a vision of the future with a possible trajectory on the premise that unless we know what we are trying to achieve it is difficult to formulate sensible policies and strategies to help us get there.

During our research into the automotive industry over the past decade or so, and particularly in the last few years, a number of trends have become clear to us, some of which have the potential to overwhelm the industry. These centre around the following two issues:

- It is becoming more and more difficult to make money making, selling and supporting cars and their components.
- Environmental and social pressures on all parts of the industry are intensifying and becoming broader in scope.

We explore the first point later on in this chapter, but our main emphasis will be on putting the industry's problems in an environmental context. In the process we will bring together a number of apparently unrelated trends, which when combined show a clear direction into which the industry is currently heading. We have found that many people in the motor industry are of necessity focused on the short to medium term 'job in hand'. Their job is to meet the next target or deadline. As a result, they often lack the overview necessary to see major trends developing. In addition, they often appear to have a somewhat distorted view of the world outside. The industry often appears to live in an imaginary world where all people buy new cars and where all new cars are sold to middle-class families with a working husband and a housewife who uses a second car for the school run and shopping at the supermarket. The great diversity of the real world often appears to pass them by; a view that reinforces the traditional attitude of losing interest in the product once it leaves the factory gates.

Visionaries do exist, even within the industry. It is significant, for instance, that the industry is increasingly showing us future concept cars which appear to address some of these problems, and we review some of these in Chapter 6. However, in so doing they often tend to move away from what much of the industry still considers its core competence: internal combustion engines and making bodies from pressed and welded sheet steel.

This is significant in that it shows in simple terms that the industry is aware of the problems, but that it seeks the solution in abandoning its core technologies. In this book we set out to analyse these 'megatrends' and the industry's response. However, we also seek to analyse how the industry could proceed to an alternative situation. In the process we redefine the core competence of the industry and provide a different interpretation of its history which goes a long way to explaining its current predicament.

The automotive industry faces an unprecedented environmental and economic crisis which calls into question the very essence of the industry. This crisis goes well beyond the prescriptive approach of the management 'gurus', who often appear to offer a simple solution to a complex problem. In the context of the car industry this is perhaps epitomised by the proponents of lean and agile production (Womack et al., 1990; Womack & Jones, 1996). On the other hand, it also moves well beyond the kind of radical accounting-based analysis of authors such as Williams et al. (1994), who although identifying many elements of the current (economic) crisis, none the less fail to identify the inherent weaknesses of the existing technology paradigm, who fail to integrate environmental issues into their analysis, and who fight shy of offering any solutions. In other words, it is not forecasting rain that is of real value; it is designing an ark that matters. This book is therefore an attempt to design an 'ark' for the automotive industry.

Although we have to spend some time in this book presenting the problems, it is important to move beyond a mere catalogue of woe. We approach these problems primarily from an environmental perspective. This book is therefore also an attempt to apply broadly environmentalist thinking to a specific industry, and to consider the implications of such environmentalism for the future structure and character of that industry. In so doing, this book makes a contribution to two debates. On the one hand, it is concerned with general notions of 'sustainability', 'environmentalism', 'whole life-cycle costs' and related concepts — where the intention is to provide an applied and empirical, substantive content to such theoretical concepts. On the other hand, it is concerned with the future of the automotive sector as a whole. The intention is to provide a reasoned, well-documented and comprehensive basis for understanding how the automotive industry may change in the future.

We do, of course, recognise that global and local environmental problems will not all be solved simply by changes to the automotive sector. Many other areas of economic activity from construction to agriculture face equally challenging futures. However, there is broad agreement that transportation in general and the automotive sector in particular represent a key area where the conflicting requirements of wealth generation and quality of life must be reconciled (Davies, 1994; JTERC, 1995; OTA, 1995; RERI, 1996). The wealth-generating capacity of the industry is considerable. Maxton and Wormald (1995: 78) make the calculation that at an average of US$20 000 per car, the value of the new car industry alone is some US$1000 billion a year. This, as they point out, makes it roughly equivalent to the total size of the economy of a country like France, the UK or Italy in 1990. To this must be added the aftermarket, parts and accessories, which roughly doubles the size of the automotive industry, as well as various related industries such as the oil industry, advertising or even insurance. Some 13% of international trade in manufactured goods involves automotive products (CCFA, 1994: 2).

The debate on the future of the automotive sector and its products has grown in intensity and breadth over recent years, as exemplified by Riley (1994), Maxton and Wormald (1995), or Cronk (1995). This is both a cause and a consequence of government policy initiatives, which are analysed in Chapter 3. Our view of what constitutes the automotive industry is a very broad one, not least because of the interrelated nature of the challenges which the different parts of the industry face. As noted above, by any measure, the automotive industry is indeed a vast area of economic endeavour, and to focus the analysis only on the vehicle manufacturers would be to miss the majority of the industry. The vehicle assemblers and their products are the most visible element of the automotive sector, which extends upstream from the assemblers through a series of overlapping supply chains of material and component producers. It also extends downstream through sales and distribution networks and the many back-up activities, from petroleum supply to insurance, advertising, and vehicle dismantling and disposal operations, which support vehicles in use. At least in the industrialised countries of the world, the automotive sector impinges upon the economic livelihood of virtually everybody.

The lifetime for the average car in an average industrialised market is a period of about 10–12 years. That is, it currently takes the average car about 10–12 years to go from vehicle assembly, to eventual disposal, dismantling, recycling and landfill (ACEA, 1994a; Brown et al., 1994). The raw materials phase would add to this a further period of up to a year. In addition, the industry requires 3–5 years on average to develop new models, which may then remain in production for between 4 and 15 years. Thus, any initiatives undertaken immediately and implemented in new vehicle designs will carry environmental implications for the next 20 years, if not longer. All of the activities associated with the life-cycle stages carry environmental implications in terms of energy and other resource use, as well as pollution.

The automotive industry has long been regarded as a key economic 'powerhouse', generating jobs and wealth directly as well as indirectly through widespread multiplier effects. In the European Union (EU), 8% of employment in manufacturing industry is in the motor industry alone (CCFA, 1994: 2). In this respect, our concern with the automotive industry is even wider because the industry is embedded within social and economic structures, and those structures will play an important part in both defining the problems the industry faces, and also defining the solutions the industry may adopt. In its broadest sense, therefore, the industry may also be thought of as a socio-technical paradigm; that is, a combination of technologies, industrial structure, cultural norms and attitudes, as well as governmental policies which collectively constitute 'motorisation'. In contemporary industrial societies, motorisation and car dependency are literally built into a spatial infrastructure which inevitably is relatively enduring.

Schumpeter (1939) and subsequent followers have argued that bundles or groups of related technological innovations and their commercial exploitation generate distinctive trajectories of development (Freeman, 1993). This school of thought involving 'long waves' of economic development has been mainly concerned with the emergence of new bundles of technology and the social innovations which accompany them. Our concern is rather different in that the focus is on how, and to what extent, an existing and established socio-technical paradigm may change fundamentally in the light of growing pressure leading to dysfunctionality and, ultimately, collapse.

In Chapter 8 we provide a broad sector and macro-economic account of the ramifications of a 'paradigm shift' in the automotive sector as a whole. Such a paradigm shift would involve a major move away from existing technologies and structures in favour of new ones. This would involve a revolution in the industry, although it would take some time to full realisation. Such a paradigm shift will inevitably be uneven — there will be winners and losers. Within the sector, the environmental and economic issues that are the subject of this book are translated into concerns with the overlapping constellations of interlinked firms which supply to the vehicle assemblers: in other words, the competitive performance of vehicle assemblers derives in no small part from their ability to manage the extended supply chains that account for the greater part of the added value in a finished vehicle. Vehicle assemblers purchase between 60% and 80% of the total cost of a car, while supply chain management has been an enduring theme for the industry from the mid-1980s (Lamming, 1994). Thus, in the context of a paradigm shift, the ability of existing vehicle assemblers to accommodate radical technological and market changes will to a significant extent depend upon their ability to create new constellations of supplier firms.

At the level of the individual firm, we relate the concept of a paradigm shift to that of core competence, a management doctrine that has received considerable support over recent years (Prahalad & Hamel, 1990; Hamel, 1991). Here the issue is one of defining what the core competence of firms in the automotive sector may be, and how that core competence will have to change in order to achieve a sustainable automotive industry (Wells, 1996a).

1.2 PERSPECTIVES ON ENVIRONMENTALISM AND SUSTAINABILITY

As debate around issues develops it becomes clouded by ambiguity and multiplicity of meaning. This is the case with environmental issues where key concepts such as 'sustainability', 'environmentalism', 'greening' and 'ecologism' have changed over time. There is a range of views that could be described as environmental, from green consumerism to radical ecologism and 'deep ecology' (Marien, 1992; Poduska et al., 1992; Schumacher, 1973). Within this range of views, our analysis could be described as radical in that we predict dramatic changes to both the products and the industry needed to build them if both are to be accommodated within a more sustainable framework. On the other hand, we paint a scenario which ensures the continued existence of a motor industry — something many radical environmentalists would not accept.

In Chapter 7 we provide a framework for the analysis of a core concept in this book, that of the environmentally optimised vehicle (EOV). In this, our starting point has to be that all human activity affects the environment in some way. What we need to do is ensure that this impact is as minimal and sustainable as possible. All human activity should be environmentally optimised — the least environmentally damaging solution, given our perceived needs in terms of quality of life and living standards, should be adopted.

However, we also want to move away from much of the environmental debate, which appears to focus on 'protecting the environment' as if it were some external entity that merits our protection. It is clear from the increasingly accepted Gaia hypothesis (Lovelock, 1979) — now adopted by the science of geophysiology (Tickell, 1996) — that the planet is a self-regulating mechanism with life as an essential part of this mechanism. However, Lovelock also argues that the earth is currently in a crisis situation, that it 'has a fever' and that the last thing we should be doing is adding more carbon dioxide: 'Sadly, what mostly shows up when we include Gaian feedbacks is that the system will amplify the damage we are causing, rather than opposing it' (Paul Valdes quoted in Tickell, 1996). As such, the planet will survive and carry on with or without us in some form. It existed long before we did, after all. In this sense, 'the environment' will never be destroyed. What is at stake is not 'the environment' as some abstract external notion, but our environment, our ability to live on this planet.

Rather than categorising the environmental movement as some vague charity, or as a pressure group to be accommodated or outmanoeuvred, the automotive industry and government will have to realise that environmental concerns are to do with self-interest. Over the past few centuries people, especially but not exclusively in western societies, have often misunderstood where their self-interest lay. As a result we have been pursuing practices which are now known to be destructive of our environment. Our primary concern must therefore be to ensure that an environment conducive to human life is sustained by seeking to minimise negative environmental impacts. In order to sustain human life, we also have to ensure the survival of other life forms, but we should not forget or disguise our inherently selfish motive in doing this (WCED, 1987).

With respect to the EOV, we develop a set of principles or parameters that would define the characteristics of the vehicle, e.g. low energy requirement in production and use, low or zero emissions, longevity, design for modularity, lightweight construction, versatility. But we also note that at the very top of the accepted hierarchy of sustainability is a 'reduction in energy and materials use' (Cooper, 1994: 11). The implication of this is clear: there must be an absolute reduction in the volume of new cars produced — something that even the most environmentally informed automotive companies appear unwilling to consider at present.

1.3   THE ENVIRONMENTALIST CRITIQUE OF THE AUTOMOTIVE INDUSTRY

The 'case' against the automotive industry from an environmentalist perspective is certainly quite strong and growing. The main areas of concern to date have been

- pollution
- congestion
- road building
- resource consumption
- scrapping, landfill and recycling
- human cost (accidents, etc.)
- wider social problems/alienation
- wider global environmental problems, e.g. global warming

Some attempts have been made to account in financial terms for the external costs of motorisation. One such study by the World Resources Institute (reported in *The Economist*, 1996) put the social costs of driving (to include road building/infrastructure, costs of illness from pollution, costs of accidents) in the USA at US$300 billion per annum, equivalent to 5.3% of GDP or US$2000 per car.

Over the past few years a number of key books have specifically sought to present the environmental problems caused by automobility, and increasingly they have attempted to offer solutions. Although warnings about the danger of automobility go back almost as far as the car itself, we will concentrate on a brief review of recent works. The recent series of studies probably started with *Rethinking the Role of the Automobile* by Michael Renner (1988) of the Worldwatch Institute in Washington, DC. Although he focuses on oil dependency and the need for greater fuel efficiency and emissions control, he widens the argument to the problems of automobility generally and the need to look for alternatives. It was followed a year later by Marcia Lowe's (1989) *The Bicycle: Vehicle for a Small Planet*, which argued imaginatively against automobility and for the bicycle as an alternative to the motor car, pointing out that in fact more passenger miles worldwide are served by bicycles than by cars and that world production of bikes outstrips that of cars by a factor of three (Lowe, 1989: 3, 13). The first half of the 1990s then saw a series of books covering topics related to cars and environment or, more widely, transport and environment. We will limit ourselves here to some works we regard as key and which deal primarily with cars and environment (Table 1.1).

Heathcote Williams' critical poem *Autogeddon* (1991) had a profound effect on the anti-car movement, particularly in the UK. However, it also made a number of 'normal' car users think for the first time about the implications of their own automobility. The poem centres around an imaginary visitor from outer space who concludes after detailed observation of our planet that the car must be the dominant life form. The second half of the book consists of a series of historical quotations on the car and its effects. The format allows a degree of emotional impact that other, more prosaic works cannot possibly achieve and as such *Autogeddon* has to be considered as one of the key works in the debate to limit our dependence on the car. It was followed in the UK by a television

Table 1.1 Key Works on Car and Environment in the 1990s

	North America	Europe
1991		Williams Zuckermann
1992		Nieuwenhuis et al. Roberts et al. (eds) Fiedler
1993	Freund & Martin Nadis & MacKenzie	Whitelegg
1994	Riley	Nieuwenhuis & Wells (eds) Berger & Servatius
1995	Cronk	Maxton & Wormald

programme, and theatre productions all over the country have been inspired by it.

Zuckermann (1991), in his book, *End of the Road: The World Car Crisis and How We Can Solve It*, was among the first to provide a comprehensive overview of the impact of the car. He looks at this from environmental and social perspectives, as well as a psychological perspective, in trying to analyse our relationship with the car.

Travel Sickness, edited by Roberts et al. (1992), takes a wider transport and environment perspective. The various contributions cover air transport and public transport as well, while taking a strong UK perspective; the book's subtitle is 'The Need for a Sustainable Transport Policy for Britain'. This can be justified by the fact that within the context of the developed countries of north-western Europe, the UK has a number of unique problems in terms of population distribution, the resulting distribution of economic activity, but also the privatisation policy and lack of government support for an integrated transport approach during successive Conservative governments since 1979. *Travel Sickness* provides a comprehensive analysis of these problems and also offers possible solutions. It was widely promoted by supporters of alternative transport modes and has been instrumental in improving the information base available to transport campaigners throughout the UK.

The Green Car Guide, by Nieuwenhuis et al. (1992), was a deliberate attempt to inform both the car-buying and car-using public, as well as policy-makers, on the full impact of cars on our environment. It deliberately moved the debate beyond the focus on emissions and air pollution only. It is wide-ranging and reflects the thorough car industry knowledge of the authors. It explains the problems of the car from an historical perspective whilst also reviewing a number of alternatives in the context of the contribution that ordinary car users can make. However, the authors also make it clear that the solutions already exist, and that to this extent no major technological breakthrough is required; rather that a more challenging change in attitudes is a prerequisite.

Fiedler's (1992) book, *Stop and Go: Wege aus dem Verkehrschaos*, is one of the German contributions to the debate. Germany, perhaps more than any other single country, owes its prosperity to the automotive industry, but also has one of the strongest 'green' political movements. Fiedler takes the problems created by the car as a given and reviews a wide range of solutions from a prescriptive transport planning perspective. However, he presents his case in an accessible manner, enabling ordinary citizens to participate in an informed way in the transport debate. He criticises the mixed signals on transport and environment coming from the German Federal Government.

Car Trouble, by Steve Nadis and James MacKenzie (1993), can almost be regarded as a US equivalent of *The Green Car Guide*. It goes beyond the traditional focus on emissions by reviewing the wider impact of cars on our environment and also suggests alternatives. It offers advice to individuals on how to modify their behaviour in order to improve the situation and is optimistic in the sense that the authors summarise the essence of the book on the front cover as: 'How new technology, clean fuels, and creative thinking can revive the auto industry and save our cities from smog and gridlock'.

Both *The Green Car Guide* and *Car Trouble* therefore sought to provide specific advice to green consumers, building on the ideas of Elkington and Hailes (1988). Nadis and MacKenzie have considerable faith in the power of consumer choice and the operation of markets to achieve a radical transformation of the automotive industry, its products, and the way they are used.

Freund and Martin (1993), in their book, *The Ecology of the Automobile*, take a slightly different view in that they focus less on the car itself and more on the impact of mass motorisation on society. They link mass motorisation directly with 'the fragmentation of social life' (p. 9) and they argue that as a transport solution, 'auto-dominated transport uses too many resources and too much energy in order to accomplish too little' (p. 23). They see the solution not in eliminating the car, but in attempting 'to exercise social control over auto use in judicious ways, to reformulate auto technology, and to diversify transport systems' (p. 24). The authors also link the growth of mass motorisation with the growth of capitalism during the 20th century and emphasise the link of the car with individualism, often at the expense of the needs of society.

John Whitelegg's (1993) *Transport for a Sustainable Future: The Case for Europe* takes a wider transport perspective, although it too focuses on the consequences of automobility, covering the usual issues such as global warming, air pollution and noise pollution. However, it adds some useful new elements to the debate by placing it in the context of sustainability and by introducing some new ideas, such as the concept of 'time pollution', for which he uses a number of older sources. Whitelegg points out that 'Since we are able to increase distances between things like hospitals, schools, and shopping centres ... but not increase the number of hours in the day, then we must increase speed' (Whitelegg, 1993: 76). Speed itself is identified as a major pollutant in that it absorbs distance and thereby 'pollutes space, time and mind ... [because] the faster the mode of transport, the more space it requires' (Whitelegg, 1993: 87). Whitelegg adds a valuable international dimension by drawing on his knowledge of Germany. This also enables him to make his political case for decentralisation as a means of restoring local democracy and hence the ability to 'act local' and improve transport at the small-scale local level where it matters.

From 1994 onwards, we begin to see publications that take a more positive view of the motor industry's ability to change. Although *Motor Vehicles in the Environment*, edited by Nieuwenhuis and Wells (1994), still tackles some of the environmental problems of motorisation and promotes alternatives such as the bicycle or light rail systems — each of which receives a dedicated chapter — it tackles these in greater detail and begins to offer some solutions such as a move towards long-life cars and how the industry could achieve this. It also has a chapter on the efforts on the part of the commercial vehicle industry to reduce pollution from its products, albeit prompted by legislative pressure. It begins to move away from addressing the basic problem of motor vehicles and environment and begins to take tentative steps to presenting a more visionary approach to the changes needed to move beyond over-reliance on automobility.

Berger and Servatius (1994) appear to take a particularly positive position in *Die Zukunft des Autos hat erst Begonnen: Oekologisches Umsteuern als Chance*. They argue that the increasing environmental pressure on the car industry to change should be seen as an opportunity rather than a threat. They argue strongly for the 'technofix' and feel that a pro-active and technologically creative approach to the problem could not only save the car industry generally, but the German car industry in particular. They argue that it is not the car itself that is the problem, but its outdated concept and the way it is used (Berger & Servatius, 1994: 71).

Although one may have doubts about their vision of a 'win–win' scenario and the primarily technological and organisational fix to the problem, the very fact that mainstream automotive consultants such as these authors dedicate a book to the problem is indicative of how seriously the motor industry is now taking the environmental debate. Berger and Servatius realise that major changes are required not only in the basic concept of the car, but also in the transport system of which it is a part and in the attitudes of car users. The authors recognise that this requires a radical rethink on the part of the car industry, drivers, politicians, planners and trade unionists, as well as traditional car lobbyists such as car associations (Berger & Servatius, 1994: 148–9). However, a close reading reveals that the authors advocate some very dramatic changes for the industry, both in terms of its products and the way it is organised.

Riley (1994) takes a more practical approach by actually developing and presenting a detailed concept of a lightweight car for the future. His book, *Alternative Cars in the 21st Century: A New Personal Transportation Paradigm*, argues that in order to improve the environmental performance of the car we should introduce a new type of vehicle half-way between a car and a motorbike. By positioning this outside the existing car concepts, Riley argues, such a vehicle can also move beyond the existing historical ballast in terms of legislation and consumer expectations that 'the car' is burdened with. He paints an imaginative picture of small lightweight commuter and city cars with a considerable 'fun' factor, drawing on his own experience of developing and marketing such vehicles in small numbers to enthusiasts. Like many US observers, he is optimistic about electric cars. One of the most useful contributions of the book is that it moves far beyond the merely technical concept and provides a realistic practical approach to how one might market alternative personal transport concepts (see also p. 198).

In a sense Cronk (1995), in his influential work *Building the E-Motive Industry: Essays and Conversations About Strategies for Creating an Electric Vehicle Industry*, follows on the debate from both Berger & Servatius (1994) and Riley (1994). Through a series of interviews with relevant experts and opinion formers in the US car industry, Cronk presents a scenario of how an electric car industry might be created by departing in a fairly radical way from the existing car industrial structure. It draws quite heavily on Amory and Hunter Lovins' 'Hypercar' concept of a very lightweight high-technology car using new space-age materials and an electric or hybrid powertrain. It is a major work within the US electric vehicle literature, but its value lies in assessing the real views of leading figures in the US car industry on the industry's future, as well as in painting a vision of that future far beyond the existing car industry structures. In this way it contributes to the same debate as Berger & Servatius (1994), which will be taken yet further in the current book.

Another significant work is *Driving over a Cliff? Business Lessons from the World's Car Industry*. Although presented as a conventional study of the car-making business, its authors, Maxton and Wormald (1995) embrace all the major issues facing the industry, such as the increasing unsustainability of mass motorisation and the crisis within the industry in terms of saturated markets, increasing competition and rising costs. As such it is probably the most useful general introduction to the motor industry in the 1990s.

We have seen then, during the early 1990s, a gradual shift from identification of the problem to a seeking for solutions, and gradually making these solutions more concrete. The danger is that as some of these ideas are adopted by the industry and become, as it were, more 'mainstream', both the arguments and the solutions will become diluted by the day-to-day economic and political pressures affecting the industry. It is important to realise that the problems have not gone away, nor has the environment improved significantly by any available measure. The longer we wait, the more radical and sudden a change will be needed.

In terms of the physical environment, it is now well established that vehicles consume vast quantities of scarce resources, produce all manner of pollutants which directly or indirectly affect human health, and are generally an extremely intrusive technology — the impact of which is compounded by the growing numbers of vehicles in use. It is not the purpose of this book to provide a detailed environmental critique of the automotive industry. Indeed, having outlined some of the key works of recent years, we take that critique as our starting point. Neither would we seek to deny that the industry has indeed improved its performance in some key areas such as emissions from engines and fuel efficiency (as shown in Chapter 5), as well as occupant safety. However, whatever the relative merit of recent improvements in vehicle design, these appear to us insufficient in relation to the scale of the problem, and the costs of achieving each marginal improvement are growing ever larger within the existing paradigm.

In neoclassical economics terms, the externalised costs of motorisation are now more able and more likely to be reflected in some form of economic cost paid by the consumer. This is the central thrust, for example, of the European Commission proposals to penalise vehicles with relatively high fuel consumption under the 'carbon tax', as shown in Chapter 3. The net effect will be to increase the economic cost of vehicle purchase, ownership and use and this will reduce demand for motorisation in general and new cars in particular. It is here then, that our line of thought becomes apparent. In Chapter 2 and later in this chapter, we show that, even without any environmental issues to concern itself with, the automotive industry as an established paradigm faces some fundamental structural problems including stagnation, fragmentation of demand, and chronic unprofitability, which will be extremely difficult for the industry to resolve. Moreover, as the environmentalist critique of the industry has become stronger over the years, and concerned with a wider range of issues, it is possible to discern three basic trends:

(a) The debate has moved outside the direct control and influence of the automotive industry (especially in terms of the development of alternative transport infrastructures).
(b) Regulation by government in the future is likely to become more holistic and systemic in character, and therefore more pervasive in its impact.
(c) The cumulative impact of greater regulation will be to reinforce existing market problems of over-capacity, unprofitability, etc., and thereby act to reinforce the pressure on the viability of the current paradigm.

In addition, the car and automobility are increasingly regarded as the cause of social disintegration. In this way, automobility is crucially involved in the way society has changed in the industrialised countries during this century. Many of these changes are increasingly regarded as undesirable and wider attempts to change the way we live could have a major impact on our need for automobility. That is, while most attention has to date been on 'top-down' environmentalism as embodied in government regulation and standards, it is possible to discern a growing 'bottom-up' environmentalism as evidenced by social movements like 'Reclaim the Streets' in the UK, and by the growing incidence of inner urban traffic bans and pedestrian zones in various countries (cf. Topp & Pharoah, 1994).

1.4   THE AUTOMOTIVE INDUSTRY RESPONSE: MEETING STANDARDS THROUGH INCREMENTAL IMPROVEMENTS

The early years of the automotive industry showed a range of competing technological alternatives for the powertrain of which the internal combustion engine was just one, while there were many divergent designs for the car body/chassis/structure, together with a large number of firms with diverse technological roots.

As Tables 1.2 and 1.3 show, many automotive companies entered into the sector not as entirely new start-ups, but as offshoots of existing companies active in other sectors; notably bicycles, other transportation products, and apparently unrelated products. A more detailed historical analysis in Chapter 9 will show various other areas from which firms moved into car making, such as armaments. In some cases the pre-existing activity was abandoned alongside or soon after the entry into automotive production. The US 'Big 3' (General Motors, Ford and Chrysler) are not included as these were all established specifically for

Table 1.2 Core Area of Existing Automotive Companies When Starting Car Production (Europe)

Make	Core area before cars
Audi-NSU	Cars (Horch)-bicycles, motorbikes
BMW	Aeroengines
Citroën	Gears
Fiat	—
Jaguar	Side-cars, coachbuilding
Mercedes-Benz	—
NedCar	Trailers, trucks (DAF)
Peugeot	General ironmongery, bicycles
Renault	—
Rover	Bicycles
Saab	Aircraft
Volvo	Ball-bearings

Table 1.3 Core Area of Existing Automotive Companies When Starting Car Production (Japan)

Make	Core area before cars
Daihatsu	Engines, three-wheel delivery vehicles
Honda	Motorbikes
Isuzu	Shipbuilding, utilities
Mazda	Cork products, machine tools, light trucks
Mitsubishi	Shipbuilding, finance, mining, chemicals
Nissan	—
Subaru	Scooters, bus coachwork, railway rolling stock
Suzuki	Motorbikes, textile machinery
Toyota	Textile machinery

car making. The Korean companies present a specific situation, which we will analyse more closely in Chapter 9.

This period in the early years of the industry was marked by relatively low entry and exit costs (see McKinstry, 1993, for the case of the early motor industry in Scotland). The new entrant of the early 1900s required very little by way of start-up capital, labour, or distribution systems. Indeed, many of the 'makes' of the early years actually made very little, as many components and subassemblies could be bought in. Few firms made their own engines, for example. By comparison, the recent entry by Samsung into the volume automotive industry required an initial investment estimated at US$2.5 billion, and a total investment of US$13 billion between 1996 and 2010 (Gadacz, 1996).

We argue in Chapter 8 that the existing cost structures in the industry are not economically sustainable. Moreover, as we show in Chapter 9, the socio-technical paradigm extends as a concept to include social and political innovations in support of motorisation, cultural norms and attitudes — and we argue that these too are changing. In its transition years into mass production, the automotive industry was seen as providing personal liberation, and vehicle production as 'modernising' the economy as a whole, while at the level of macro-economic policy there was a supportive regime under Keynesian economics. This post-1945 period was characterised by consumer-based expansion of material wealth, and a social contract between capital and labour epitomised by the automotive industry (Aglietta, 1979).

All these features came together to make conditions right for the continued development of the industry along its trajectory of high fixed cost and high volume output, allowing the further development of the internal combustion engine and the all-steel body. However, these conditions are not immutable, inevitable or fixed. On the contrary, in terms of the socio-technical paradigm argument, external conditions are now — in its mature phase — turning against the industry. If history is any guide, this may lead to the gradual (or even sudden) decay and displacement of the industry, as has happened with other seemingly established industries in the past.

Within the automotive industry the socio-technical paradigm is embodied in two core product technologies: the all-steel body (Chapter 4) and the internal combustion engine (Chapter 5). In each case these two interrelated product technologies in turn derived from important related innovations in materials and process technologies, and in due course became established as the dominant technologies for the mass manufacture of vehicles. Chapters 4 and 5 explain how this occurred, and more importantly the consequences for the basic economics of the sector until today. One vital feature of the basic set of economic rules is that entry and exit costs have risen inexorably. It is not cheap to become a volume car producer, nor to decide to stop being one.

The practical application of environmental concerns over the automotive industry has, historically, been quite limited. In terms of governmental regulatory policies there have been two broad categories of intervention: those related to the product, and those related to use. Technical 'quick-fix' and (quite literally) 'tailpipe' solutions have been available to the industry, albeit at some cost. That is, the industry has been able to continue to make relatively small incremental improvements in the product and processes used to create it, while regulation has been defined by a fairly minimalist view of the car's environmental impact and consequently of what actions need to be taken. The industry to date has been able to react to and contain environmentalist pressures but will be less able to do so in the future.

Examples of recent incremental improvements in the automotive industry could include

- water-based paints
- exhaust catalysts
- side impact bars

- engine management systems
- reduced weight/complexity of the all-steel body
- low rolling resistance tyres
- intelligent vehicle highway systems (IVHS)

Thus, for example, IVHS can be seen as a means to squeeze extra capacity out of the existing road infrastructure, and make a marginal contribution to improved vehicle efficiency by reducing congestion. A more cynical view might be that such innovations merely result in increased road traffic, and to this extent are counter-productive.

The automotive industry continues to improve its product and its production processes in all sorts of ways. New developments such as Nissan's body-framing technology allow a greater mix of model variants to be processed down one production line, for example (Abe et al., 1995), as do developments in assembly fixtures (Keebler, 1995). Laser-tailored steel sheets offer better material yield and allow closer optimisation of steel types for the application in question (Ujihara & Cooke, 1992; Baysore, 1995; Hanicke, 1995; Hoeven et al., 1995; Moerman & de Bleser, 1995). In terms of powertrain we have seen the development for mass production of multivalve engines, turbocharging, engine management systems and more sophisticated transmissions. That is, the automotive industry has continued to refine the basic paradigm over a long period of time in an incremental manner.

At issue here is whether further improvements in the paradigm are sufficient, and whether the automotive industry has the potential to move beyond incremental improvements towards more radical solutions. In Chapter 6 we examine the extent to which the automotive industry has thus far generated truly radical solutions. In Chapter 7 we take this further to suggest — largely based on available technologies — the new technological trajectory which the industry and legislators may take if incrementalism proves insufficient to the task of meeting the economic and environmental crisis which the industry faces. In Chapter 8 we follow through this argument to look at the impact such a radical change might have on the composition of the automotive industry; while in Chapter 9 we attempt a long-range view of the automotive industry in the year 2020.

1.5 REGULATION AND COLLABORATION: GOVERNMENT AND INDUSTRY IN AN ENVIRONMENTALIST WORLD

The hand of government is and will remain a critical feature of the competitive context within which the automotive industry exists. Our case here is that regulation and collaborative R&D programmes will go hand in hand (essentially a 'carrot and stick' approach), as we show in Chapter 3. However, the character of government intervention is changing, from limited and essentially punitive control, towards a much more pro-active and collaborative stance (Maxwell, 1995).

While collaborative R&D programmes, such as the Partnership for a New Generation of Vehicles (PNGV) in the US, are certainly flawed and, possibly, ineffective or only partially effective in terms of resolving the environmental–structural crisis that the industry faces, in this book we argue that the combination of such R&D programmes with new regulatory instruments will have important effects at all levels in the automotive industry.

Our EOV (Chapter 7) is a market concept, yet it is also a production concept (that is, we expect it to entail new approaches, technologies and materials in production). However, it is first and foremost a regulatory concept, in that the EOV reflects a much broader approach to defining and moving towards sustainability in the automotive sector. It is too early to claim that regulation has reached such a holistic and whole-life-cycle approach, but it is certainly moving in that direction.

At the level of the individual firm the problem is one of defining the core competence of the firm. Chapter 8, on the economic implications of the EOV, touches on this theme, but this issue also underpins the material in Chapter 2, which deals with markets: do vehicle assemblers make cars and sell them, or do they have a broader brief, for example as systems integrators or as providers of personal mobility services?

1.6   THE OTHER PROBLEM: LACK OF PROFITABILITY

The public perception of the automotive industry is still one of a sector that generates large amounts of profit. If this was ever true it certainly has not been the case for the past few decades. As we indicated at the start of this chapter, lack of profitability is one of the key factors in prompting a change of strategy in the industry. Absolute profit figures in the industry look large compared to other sectors, dominated by small or medium-sized businesses. By its very nature, volume car production in a mass production system requires large businesses. However, in terms of profit margins, most players in the industry are barely profitable. The short periods of high demand generate modest profits, which are largely absorbed by product development as well as providing a cushion to ride out the next lean period, when operating losses are common. Williams et al. (1994: 76) point out that:

> The depreciation charge is relatively fixed, and so if labour's share is high and rising because of cyclical and other pressures, net income (retained and distributed profit) becomes a precarious residual. Net income is a source of cash, the existence of which is always threatened by the prior claims of labour and depreciation ... The basic point is that the surplus in the form of profits pre-tax or net income may be substantial in good years but in a number of cases negative or non-existent in bad years.

Williams et al. (1994: 77) illustrate the point with data on the good and bad years for a number of major car makers, as illustrated in Table 1.4. The point can

Table 1.4 High and Low Profits to Sales 1980–1991 for 12 Major
Car Manufacturers (Gross Profits as a Percentage of Sales)

Firm	Highest return	Year	Lowest return	Year
Toyota	6.0	1985	3.7	1982
Nissan	3.5	1981	0.5	1987
Honda	10.2	1981	3.1	1991
Mazda	1.9	1985	1.2	1989
GM	7.2	1983	−3.2	1991
Ford	11.1	1988	−2.1	1980
Chrysler	12.9	1985	−11.1	1980
VAG	8.0	1985	0.6	1982
BMW	9.6	1984	3.1	1991
Fiat	9.0	1987	5.5	1991
PSA	12.9	1990	0.8	1986

Source: Williams et al. (1994: 77).

be further illustrated with some results from the mid-1990s, a period when the main economies were not in recession. Renault made a comeback during the early half of the 1990s with an updated production system, improved quality and an imaginative product range. It also enjoyed a home market boosted by government-led scrappage incentives, which generated new demand for small cars of the type for which Renault has an established reputation. Nevertheless, Renault returned to an operating loss during 1996.

One of the main causes lies in the decline in Renault's overall production volumes from 1.65 million cars in 1989 to 1.52 million in 1995. Break-even points in production, combined with the unrelenting pressure of product development costs, are such in the motor industry that even a small decline like this can turn a modest profit into a massive loss. However, price increases are not the answer as there is a widespread belief in France — and elsewhere — that car prices are already too high and Renault is committed to an average cut in price of FF3000 between 1996 and 1998 (Genet, 1996). Renault shares, worth FF165 each at the time of the first tranche of privatisation in June 1994, had lost 31% of their value by August 1996.

Fiat too, after a few profitable years, made a loss for 1996 (Way, 1996), while Volkswagen (VW) was in celebratory mood after a profitable year in 1996, making a 1.3% return on sales (Kurylko, 1996), thus illustrating the thin line between profit and loss in the motor industry. Its forecasts to boost this to 8% by 2001 sound optimistic in this context. VW made losses of DM1.9 billion in 1993, compared with a profit of DM336 million in 1995.

Although the US Big 3 announced combined profits of US$4.8 billion for the second quarter of 1996, this came only two years after General Motors (GM) posted the largest loss in corporate history, and it hides a number of threats in this, the most successful car market. For example, GM managed to turn itself around from the massive losses of the early 1990s and posted a profit of a record US$6.9 billion for 1995; however, this figure represents only a 4.5% return on sales of US$152.6 billion, which GM does not consider adequate. Ford recorded after-tax profits of US$4.1 billion for 1995, down 22% from the year before. However, of this about US$2.1 billion came from its financial services, which even outperformed American Express. This also hides the final quarter decline to US$660 million compared with US$1.6 billion for the same period in 1994. In contrast to its financial services, Ford's US automotive operations recorded quarterly profits of only US$168 million, compared with US$745 million the year before, while international businesses lost US$152 million. The avowed aim of the cost-reduction and corporate restructuring plan that is Ford 2000 is to raise average annual profitability to just 5% of turnover.

Nissan, based in Japan, the home of cost saving and lean production, made its fourth consecutive loss in 1995, the equivalent of £526 million (Feast, 1996). Yet Nissan is widely regarded as one of the more successful car producers worldwide. Williams et al. (1994: 41) make a comparison of average profit per vehicle sold as a percentage of average sales revenue between Nissan and VAG, makers of Audi and Volkswagen cars, which shows not only the very small profits generated even by successful companies, but also the perilous state of a firm such as VAG even during the boom years of the 1980s (Table 1.5).

Recovering product development costs on such a modest return is not easy. Volkswagen's Passat, never its best-selling model, cost between £500 million and £1 billion to develop (Williams et al., 1994: 41). Product development has to continue even during lean years as good attractive products are the key to success. It is indicative that the Ford 2000 programme aims to reduce development costs by US$1 billion per annum, while Ford's investments in new products and facilities in 1995 were US$8.7 billion.

Table 1.5 Average Profit per Vehicle as a Percentage of Average Sales Revenue per Vehicle Sold, 1981–90

Year	Nissan	VAG
1981	3.5	−0.7
1982	3.1	−0.5
1983	2.6	0.5
1984	1.7	1.1
1985	1.8	1.1
1986	0.6	1.1
1987	0.5	1.0
1988	1.5	1.2
1989	2.4	1.6
1990	2.0	1.4
Average	1.97	0.78

Source: Williams et al. (1994: 41).

Recent requirements of environmental and safety legislation have added considerably to the already high cost of product development. At the same time, the Japanese drive to 'catch up' with the West in terms of product has forced a general shortening of model cycles, although few producers have achieved the four-year replacement cycle many Japanese producers use for their more popular models (Chapter 8). In fact, even the Japanese are now beginning to favour longer cycles in an attempt to control costs. Nevertheless the pressure remains to replace models more often than in the past. It is clear then that despite the popular image, car making is not a profitable business. However, as we shall explore in the course of this book, much of this is due to the technology the industry has adopted for building its cars' bodies. This technology is amenable to change, albeit at considerable cost. However, the ultimate prize may be a profitable and more sustainable industry.

In the meantime the industry is attempting to recover the economies of scale needed for profitable operations through globalisation. It is argued that product development costs can be dramatically reduced for global players by developing one single product range for all world markets. This will lead not only to a reduction in product development costs but also to economies of scale by making larger numbers of each model. In practice the true 'world car' has eluded most car makers in the past. Ford's US Escort of the 1980s shared only a handful of parts with its European counterpart, despite both starting off as a single project. Ford's efforts on Mondeo (Europe) and Contour/Mercury Mystique (North America) were more successful, although both cars still manage to look different. Under the Ford 2000 programme this globalisation process will be taken even further and this approach will dominate the industry over the next 15–20 years. However, ultimately the basic constraints of the way volume cars are made will once again come to the fore. It is possible that the first to address alternative ways of doing things are those producers that do not share the ability of Ford and GM to go global. As a result it could well be some of the smaller European and Japanese players that will break the mould.

Moreover, profitability problems are not confined to the vehicle manufacturers alone. Components and materials producers face repeated impositions of price reductions from the vehicle manufacturers. Profit margins on original equipment sales, traditionally quite low for high-volume components, used to be offset to no small degree by the higher margins enjoyed in the parts replacement market (or aftermarket). However, margins here have also been under threat, while greater component longevity has eroded the volume of business faster than a growing car parc has been able to expand it. One result has been widespread, large-scale restructuring of the components industry of the type typified by the 1996 merger of Lucas with Varity. Equally, the franchised dealership networks used by the vehicle manufacturers to sell and support their cars are struggling with chronic low profitability.

The widespread restructuring in the industry is also having its effect on jobs. In 1993 the EU automotive sector lost some 70 000 jobs, while it has been forecast by the Boston Consultancy Group that around 400 000 jobs will go in the supply industry alone during the 1990s (Dillen, 1994; Schampers, 1994). The importance of the sector as a provider of employment could thus gradually be eroded.

1.7   A FUTURE FOR THE AUTOMOTIVE INDUSTRY?

Having made our case that current automotive technologies (product and process) are facing increasing pressure in terms of their ability to generate consistent profits, we move on in subsequent chapters to explore its problems in terms of (a) profitability in the face of volatile markets; and (b) environmental regulation which is redefining those markets. We then go on to argue that the overall context within which the industry works is changing. With this come new 'terms of competition', which have an impact across all aspects of the automotive industry.

With further enhancement and refinement of existing technologies and processes, the automotive industry may be able to stave off the evil day, but still the basic case remains: there is going to have to be a quantum leap in performance in order to meet the targets of, for example, the US Partnership for a New Generation of Vehicles (PNGV), which will almost certainly necessitate new materials, technologies and strategies.

Of critical importance is the pace of change. How quickly will this paradigm shift occur? Will it occur in the same way, and at the same time, in all places? Which companies and locations will lead the process of change? Alternatively, are there obvious likely laggards in the change process? Chapter 9 seeks answers to these and related questions, notwithstanding the large degree of uncertainty in making such predictions. In so doing, we hope to make a useful contribution to what is an increasingly important and urgent debate.

2 Markets and the Future of Car Technology

A number of structural changes are taking place within the mature markets for passenger cars. Most notable of these is the level of market saturation that has now been achieved and which constrains further new growth in many countries. In the medium term, the anticipated growth in legislation, higher taxes and restrictions on use will make car ownership increasingly less attractive than today.

(Maxton & Wormald, 1995: 156)

2.1 INTRODUCTION: MARKET CONDITIONS IN THE 1990s

The future of vehicle technologies will depend on the size and composition of the global market, and especially the major regional markets of Europe, North America and Japan — the so-called Triad (Ohmae, 1985). While there is thought to be considerable growth potential outside those regions, there are significant uncertainties which make prediction suspect. While regions such as China, Southeast Asia, India, Eastern Europe and South America may have potential for growth in sales, they show little prospect of being the source of technical change in the automotive industry. The only potential exception here is South Korea, which is building an automotive industry capable of independent product innovation.

It may well be the case that regions outside the 'Triad' have different automotive needs and priorities. To date, few vehicles have been engineered specifically to meet those needs: an exception here is the Palio, a small car launched by Fiat in 1996 to be built and sold within markets outside the Triad and in volumes near to one million units per year. In Korea there have been few attempts at introducing novel technologies, though Kia has built a trial aluminium version of one of its models and has launched its updated version of the composite Lotus Elan. A number of manufacturers have experimented with electric vehicles, mainly under the aegis of the Korea Automotive Technology Institute (KATECH). The relative sizes of the main regional markets are shown in Table 2.1.

It is the three major markets of the Triad that will be the prime location for innovative product introductions onto the market, and the prime source of those innovations. There is a growing trend for other regions to adopt near-

Table 2.1 Global Vehicle Sales, 1995 (000s)

Region	Cars	Commercial vehicles	Total
Africa	435	229	664
Asia-Pacific	7 987	5 114	13 101
Central/South America	645	2 252	2 898
Eastern Europe	1 603	360	1 964
Middle East	747	351	1 099
North America	9 425	6 663	16 088
Western Europe	12 006	1 193	13 200
Other	79	45	124
Total	32 931	16 209	49 140

Note: totals may not sum due to rounding.
Source: adapted from Anon. (1996a).

contemporary technology, so that the time lag as technology is diffused to new locations is probably shorter than used to be the case. China, for example, was offered some very up-to-date concepts, such as a version of the Mercedes A-class, under the tenders for its 'popular car' project. Thus new material and body concepts, and new approaches to powertrain, may be spread quite rapidly through global corporate networks. Equally, global car concepts such as the Ford Mondeo/Contour are becoming more prevalent. Whether this is evidence of markets around the world becoming more homogeneous is debatable.

As we will see later on, market growth outside the Triad regions may be important in sustaining the existing technology of welded steel automotive body structures, because market expansion reinforces the value of economies of scale in the steel unitary body. This is the 'global niche' concept (often associated with Mazda) in which a product is designed to fit a niche within each national/regional market, but where global sales give sufficient overall volume. If only built for one national market, the costs of a steel unitary body would be too high. It is also the case that expansion of the market in volume terms is attractive for any model of vehicle and would thus reinforce the predominance of steel bodies.

The situation for engines is somewhat different as these are less dedicated and can be made to fit a range of different bodies and 'platforms'. A platform is a somewhat elastic concept in the automotive industry, but in simple terms consists of the main structural elements of a vehicle body such as the floorpan and bulkhead, together with the associated suspension points, engine mountings and other fixed points. In this sense, the high investments in engine or transmission plants are more economically viable than the high investments in body technology. We will analyse these issues in more detail in Chapters 4 and 8.

The global passenger car registrations given in Table 2.2 illustrate how difficult market conditions have been in recent years, following a period of expansion in the 1980s. The market in North America remains large but highly

Table 2.2 Car Registrations, Japan, US, Europe, 1985–94 (000s)

Year	US	Japan	Europe (12)
1985	11 045	3 104	9 584
1986	11 463	3 146	10 534
1987	10 225	3 274	11 291
1988	10 569	3 717	11 879
1989	9 771	4 403	11 320
1990	9 295	5 102	12 193
1991	8 176	4 868	11 825
1992	8 210	4 454	12 614
1993	8 515	4 199	10 672
1994	8 991	4 210	11 031
1995	8 636	4 443	12 079

Source: L'Argus (1995), Anon. (1996a).

cyclical — a feature which offers two interpretations. On the one hand, the size of the market encourages large production volumes, and hence a greater inclination towards steel vehicle body structures. On the other hand, the large swings in market demand make the high fixed costs of steel production technology a real liability.

The market in the European Union (EU) is often thought of as the 'largest in the world' but in fact is more fragmented than might be supposed. While there has been considerable progress in terms of technical harmonisation of component and vehicle type approval systems and other legislative aspects of the Single European Market, wide differences remain in acquisition and usage tax systems, vehicle periodic testing regimes, topography, weather and climate, social structure and culture. These differences tend to shape the market, and producers tend to design products primarily for their domestic market and culture.

The European market has also shown a marked stagnation after the early 1990s. This is despite the introduction of scrapping incentives on older vehicles in several markets, notably in France where three such schemes have been introduced. The first scheme in France boosted annual sales by an estimated 250 000 cars in a market of around two million — a significant short-term effect. However, in general terms the European market is both saturated and very competitive. From 1999, Japanese imports will have unrestricted access, while imports from Korea may be expected to grow substantially. Significant volume growth beyond the peak years of 1989 and 1990 in the European market is unlikely before the year 2000, and uncertain thereafter (see AID, 1995, for some projections for the market in Germany, for example).

The market in Japan peaked in 1990 with just over five million registrations, but fell by nearly 20% of this total by 1993 and 1994. This consistent and large decline in the domestic market has come as something of a shock to the domestic producers, who responded with even more intensive efforts to drive down costs. More drastic restructuring of the industry has often been mooted, though to date has not materialised. On the other hand, General Motors has taken management control at Isuzu, Nissan is taking a more direct interest in Subaru, while Ford has taken effective control of Mazda. This restructuring has been accompanied by a relocation of production facilities outside Japan, by both the vehicle manufacturers and the components industry, with Honda having the largest proportion of its production outside Japan. In addition there has been a marked increase in the number of product exchanges between the different producers. Honda's domestic range is made up of the products from three competitors in addition to its own products, for example.

The market in Japan, prior to its early 1990s collapse, had undergone a prolonged period of relatively stable growth — a feature which nurtured the development of the Toyota Production System and allowed manufacturers to pursue short model lifetimes. The latter was part of a MITI-led policy of trying to allow the Japanese car producers to catch up in terms of product technology with Europe and North America — something they have now achieved. Additionally, export markets were used to expand overall sales and, as importantly, absorb any fluctuations in domestic demand. By the early 1990s, Japanese producers faced rapidly rising domestic production costs, a seemingly unstoppable rise in the relative value of the Yen, and a dramatic decline in the size of the domestic market. Sales in Japan in 1994 showed only a 0.3% growth over 1993 at 4.2 million cars, while 1995 sales were only 4.4 million.

Imports to the Japanese market have risen too, though by the mid-1990s they still accounted for under 300 000 of total sales each year. Exports in 1994 fell by 14% compared with 1993, to 3.3 million units, continuing a long-term trend established in 1986. An important feature of the Japanese market is the relatively short average life of vehicles in the parc. This is at least partly the result of the stringent vehicle testing regime which results in vehicles being scrapped or, as is frequently the case, exported as used vehicles to less demanding markets around the world. Some relaxation of this in 1995 may help change the make-up of the parc and may have a dampening effect on new car sales. Overall, vehicle sales are unlikely to recover to the peaks of the late 1980s until at least the year 2000, while exports too will decline as 'transplant' production around the world is stepped up.

2.2   INTER-MARKET PENETRATION AND THE LOSS OF PROTECTED MARKETS

In two of the major regional markets, North America and Europe, a key issue has been the loss of domestic market share or, alternatively, the increasing proportion of total sales accounted for by sales outside the domestic market. Table 2.3 shows, for firms in the EU, the proportion of total sales accounted for by domestic sales. This indicates that domestic sales have become less dominant

Table 2.3 Domestic Sales as a Proportion (%) of Total Global Passenger Car Sales, 1984–93

Firm	1984	1985	1986	1987	1988	1989	1990	1991	1992	1993
Fiat (Gp)	65.04	65.97	63.29	64.29	65.32	65.06	62.24	58.21	59.89	56.75
PSA	40.04	41.98	39.00	40.52	39.53	37.90	37.99	35.83	33.62	30.52
VAG	30.64	27.88	29.23	31.22	28.52	26.87	26.48	34.90	34.52	32.98
Renault	35.70	34.36	39.72	40.55	40.92	40.05	41.53	33.75	35.94	34.46
Alfa	60.15	65.37	60.76	60.06	64.23	64.55	62.16	58.27	66.51	64.70
Audi	41.30	37.05	42.73	51.63	45.63	39.23	38.87	48.92	48.00	47.05
Citroën	46.10	44.86	41.88	40.41	37.48	37.14	37.19	33.77	33.32	31.29
BMW	37.31	32.52	33.33	32.61	37.69	37.64	37.08	42.15	40.87	37.90
Fiat	64.31	63.86	61.96	62.48	63.05	62.48	59.68	54.73	56.26	52.77
Lancia	74.66	77.97	77.18	77.27	79.16	79.10	74.75	75.09	76.89	77.43
Mercedes	45.74	49.57	50.28	47.72	47.36	44.89	44.97	50.09	47.90	46.42
Peugeot	37.00	40.45	37.53	40.59	40.74	38.35	38.43	36.75	33.79	30.06
Rover	77.48	70.39	71.81	63.60	69.68	66.70	60.53	57.75	56.72	58.57
Saab	29.02	24.00	23.16	22.52	22.47	24.90	21.97	22.03	16.42	18.50
Seat	41.47	32.12	36.99	40.36	36.14	34.69	30.31	26.84	29.61	29.79
VW	31.76	28.91	29.60	30.55	28.55	27.50	28.02	33.13	33.17	31.87
Volvo	16.66	18.01	15.65	16.80	14.82	12.87	10.33	10.81	11.94	10.25

Source: adapted from Wards (1994).

Table 2.4 European Union Sales as a Proportion (%) of Total Sales, 1984–93

Firm	1984	1985	1986	1987	1988	1989	1990	1991	1992	1993
Fiat (Gp)	90.86	89.94	89.98	92.46	93.27	93.23	91.97	89.80	89.47	83.19
PSA	76.52	80.21	80.22	82.87	83.67	82.43	81.24	82.94	83.49	80.46
VAG	56.62	55.28	54.90	60.70	61.25	62.11	60.88	63.71	65.84	62.67
Renault	70.86	74.57	78.74	80.82	81.51	82.15	82.00	81.81	80.00	75.53
Alfa	85.95	87.91	87.36	91.04	92.88	92.22	92.06	92.38	93.10	93.16
Audi	66.26	62.80	67.19	76.49	80.07	79.60	75.25	77.47	76.90	80.70
Citroën	93.04	91.40	90.93	85.36	84.05	84.42	84.45	82.77	89.87	85.22
BMW	64.52	59.09	60.75	58.85	67.07	68.75	65.20	70.25	69.65	65.40
Fiat	90.95	89.32	89.42	91.94	92.70	92.62	90.96	87.86	87.75	80.05
Lancia	95.53	95.00	95.35	96.35	96.83	97.17	96.78	97.44	97.20	96.94
Mercedes	60.69	66.14	66.78	66.64	68.49	69.48	68.63	73.28	72.75	70.30
Peugeot	68.22	74.27	74.74	81.38	83.45	81.26	79.46	83.04	79.79	77.61
Rover	97.50	89.44	98.02	87.14	93.02	87.42	83.30	98.13	88.65	88.54
Saab	22.32	19.12	21.68	21.32	24.49	30.37	35.24	35.04	38.04	38.14
Seat	73.46	69.86	81.95	92.80	88.76	90.53	92.04	93.70	97.36	94.89
VW	52.57	52.35	50.40	53.83	53.91	54.85	53.93	56.79	59.40	55.48
Volvo	63.66	64.73	64.87	62.02	53.11	51.40	50.55	56.71	65.95	53.30
Ford	31.19	29.13	30.83	33.64	31.37	33.61	35.77	40.52	37.89	32.42
GM	15.57	15.95	16.83	20.51	20.61	23.58	24.93	28.12	29.06	25.26

Source: adapted from Wards (1994).

in recent years, but Table 2.4 shows that the EU industry is still largely dependent upon the EU market for sales.

Between 1968 and 1989 the import share of total sales grew from 21.4% to 38.1% in France; from 20.4% to 30.2% in Germany; from 8.3% to 57.0% in the UK; from 15.2% to 42.2% in Italy; and from 59.6% to 74.5% in Sweden (Wells & Rawlinson, 1994a: 39).

Exports from the EU are largely the preserve of the specialist producers such as Mercedes and Volvo, with the US as the prime export market. Of course, one of the features of the shift from national markets is that the notion of a 'domestic' market is becoming more difficult to sustain as production facilities become more geographically dispersed. The US firms, Ford and GM, illustrate the problem. Ford first established manufacturing operations in the UK in 1911 and built its Dagenham plant in 1931. Ford has the largest share of the UK market, and is widely regarded as a 'domestic' producer, just as the GM-owned Vauxhall company is. Moreover, both Ford and GM have major manufacturing operations in several European countries and in this sense have long been 'multi-domestic'. As the vehicle manufacturers move towards global vehicle development and production, the notion of 'domestic' will carry less and less significance.

The vehicle manufacturers in the EU had, for much of the post-1945 era, enjoyed growing domestic markets and the protection of import tariffs and quotas. Protectionism was matched by state ownership of 'national champions' typified by the volume producers such as Renault, VW and British Leyland (Bhaskar, 1979). However, with the creation of the European Economic Community and its enlargement into the European Union, quantitative and qualitative restrictions on trade between the Member States have been largely removed (although important vehicle taxation differences remain). This has both allowed and encouraged the interpenetration of formerly distinct national markets within Europe.

The case in North America is rather simpler and better known, largely because of the persistent growth in the share of the market accounted for by Japanese producers, initially through imports and, as the 1980s progressed, through production facilities located in that region. Between 1968 and 1989, the share of the US market accounted for by imports grew from 10.5% to 35.9% (Wells & Rawlinson, 1994a: 39).

In the case of Japan, imports had traditionally been excluded by a combination of technical and qualitative barriers, never getting beyond 2% of the market until the mid-1980s. More recently there has been some growth in imports from two sources: Japanese producers exporting vehicles to Japan from their production facilities elsewhere in the world; and imports from other vehicle manufacturers, notably VW.

The critical point to note about this process is that the former protected national or domestic markets offered a degree of certainty for the vehicle manufacturers. These markets could be reliably expected to consume the bulk of production and there was a high degree of loyalty to the domestic producer. For volume producers and their main products, those national markets remain fundamental to the economic success of any particular model, but these markets are now rather less certain — a factor which compounds the risks associated with the investments required to produce a new model.

2.3   STRUCTURAL MARKET CHANGES: SATURATION

Since 1945, US, European and then Japanese car makers have become used to an ever-growing market. Demand for cars during the 1950s and 1960s was such that manufacturers could sell virtually anything they built. This may have led on the one hand to occasionally far from perfect products; however, the customer was also provided with a great choice of technically interesting cars. The engineers were allowed considerable creativity, which led to such landmark vehicles as the Citroën DS and the Mini.

It is only really from the 1970s onwards that competition began to bite and car makers began to consider the fact that they needed to build cars which not only excelled in one specific area, but that were good all round. The politically driven energy crises of 1974 and 1979 were major causes of this trend. However, although competition increased and the products became more conservative as a result, car manufacturers and their franchised dealers in Europe still enjoyed rapid growth in new car registrations during the 1980s.

In the early 1990s, as general economic conditions deteriorated in successive national economies, so a widespread and unusually rapid decline in new car registrations was experienced. Despite this decline, the overall prognosis was that demand would recover as national economies lifted out of the trough of the economic cycle and growth resumed. The industry and its advisors were still not prepared to 'think the unthinkable', that like North America, the EU and Japan could well become markets of essentially zero overall growth.

The recovery in demand for new cars has remained elusive. Perhaps the steep decline in demand experienced as economic recession set in should have been interpreted more carefully as a clear warning that more fundamental shifts in the character of that demand were under way. That is, there is growing evidence that European markets in particular are saturated — even over-saturated — to such an extent that the level of new car registrations growth of the 1980s will never appear again. Indeed, we consider there is a very real prospect of a plateau or even continued decline in demand for new cars. A combination of economic, social and political changes may be fundamentally transforming the market (Wells et al., 1996a). In addition, the technical improvements implemented throughout the 1980s have led to more long-lived cars, thus further depressing new car demand (cf. Nieuwenhuis & Wells, 1994: chapter 10).

The most immediate and obvious reason for stagnation in the European car market is that prevailing economic conditions have been far less favourable than in the 1980s. While economic growth tends to be cyclical in nature, the recessionary conditions of the early 1990s were unusually severe in several key

Table 2.5 GDP per Head and Car Ownership Rates in Selected Countries

Country	Per capita volume index for GDP	Number of people per car
Italy	75	1.9
Japan	84	2.9
USA	100	1.7
Netherlands	71	2.7
Belgium	76	2.5
UK	70	2.3
Germany	88	2.0

Note: this index is relative to the US, which is at 100.
Sources: OECD, SMMT.

markets, including Germany, France and, particularly, the UK. Traditionally, economic growth and growth in new car registrations have gone hand-in-hand. Indeed, the poor economic performance of key national economies in Europe in the 1990s has been an important contributory factor in the decline in demand for new cars. Yet, if the 1980s ushered in an era of 'jobless growth' then the 1990s may mark an era of 'car-less growth'. That is, there appears to be a deepening separation between growth in the economy as a whole, and registrations of new cars.

It is already well known that different national markets do not show equivalent vehicle ownership rates. So while there may be a generally assumed relationship between per capita income and car ownership rates per thousand of the population, there are clear discrepancies which suggest that other factors can play an important role in shaping demand for cars overall and new cars in particular (Table 2.5).

One reason for the separation between aggregate economic growth and new car demand may be the much-discussed phenomenon of consumer confidence — especially in the UK, but increasingly in other economies such as Italy, Germany and France where social unrest is evident over broad changes in government policy towards a more monetarist position.

The UK car market is distorted by the very high proportion of company-car purchases, which have sustained at least some growth over recent years. Over time this company-car market will be reduced to essential users only, as subsidised car ownership and use is not sustainable in the longer term. Already there has been a growth in the number of companies offering a cash alternative to company cars. As a result the UK market, like many others in Europe, will come to rely more on private buyers, most of whom currently buy used cars. Private buyers are less willing than companies to carry the burden of high new-car prices and high depreciation. As a result even the remaining new-car buyers will be more price sensitive. In addition, the decline in company car purchases will contribute to a general 'downsizing' and less of a willingness to pay a premium for a prestige badge. At present in the UK most new specialist cars (Mercedes, BMW, Audi, Volvo, Saab, Jaguar) are sold to company buyers rather than private buyers.

However, it is the private buyer who is the one most lacking in confidence. In many of the advanced industrial economies there are structural shifts in working conditions, with a declining proportion of the total population being economically active, and with a greater proportion of jobs being temporary or part-time. A secondary feature in the economic basis for new-car demand may be that of the distribution of income. The distribution of income issue has several elements. First, those in employment have seen real disposable income increase relative to those not in employment. But, second, much of that relative increase is due to intensified working conditions (especially overtime) in the context of the hightened uncertainty over job security. In Japan, the 'salaryman' class of middle-class managers and bureaucrats — and the economic basis of mass consumption — has begun to be eroded by similar social changes.

A further feature of this income distribution issue is that of the ageing population. It is illustrative that Rover's MGF, ostensibly aimed at younger people, had a waiting list where the average age of buyers was nearly 60! A similar age profile occurred with Twingo buyers in France, despite the 'young things' image of the car. Although at present many elderly people have considerable disposable incomes and as such provide an opportunity, the size of the elderly population is growing rapidly over the next few decades, putting greater pressure on resources. The current debate on the future of many aspects of the welfare state, especially health care, is adding to the insecurity people in middle age feel about their future. Few can afford to secure a completely private provision for their old age or possible ill health, and the remaining majority will expect the state to continue to provide. The growing cost to the state for providing this security, however limited, will inevitably lead to higher taxes and as a result a decline in disposable incomes over the next few decades (Maxton & Wormald, 1995: 175–6).

Further complications include the greater proportion of female buyers, who again may have different priorities from traditional male car buyers, and the fragmentation of households as a result of the rising divorce rate and other social changes which are rendering the traditional 'family' car an object of declining relevance.

Demand for new cars is clearly related to the price of those cars. In other words, cars are price elastic. There is some evidence to suggest that new-car prices in Europe are too high — though any price data need to be treated with caution in the context of a market which features widespread discounting, low-cost financing including personal leasing programmes, trade-ins and other features in the exchange between a vehicle manufacturer (via the franchised dealership) and the consumer. In many ways, the differences in real price between the USA and Europe which have existed since the early 1900s are exemplified by the marketing strategy for people-carriers and off-road vehicles. In the USA,

Table 2.6 Relative Prices of Cars in the UK and USA (£), 1996

Model	US price	UK price
Hyundai Accent	5 267	8 999
Ford Contour/Mondeo 2.0	8 987	13 185
Honda Accord 2.0	9 845	14 405
Ford Probe 16V	9 082	17 165
VW Jetta/Vento 2.0GL	9 291	14 960
Mazda MX-5 1.8iS	11 667	17 595
Honda Prelude 2.3	13 137	20 895
Saab 900 S 2.3 three-door	15 644	17 995
BMW 318I	16 691	17 220
Audi A4 2.8/2.6	17 278	22 115
Volvo 850 SE20V 2.5 estate	17 881	22 300
Mercedes C220/Elegance	20 179	26 000
Lexus LS400	34 490	45 995
Honda/Acura NSX	54 116	68 245

Sources: derived from Road & Track (1995); Complete Car Jan. 96. Exchange rate: US$1 = £0.652.

prices are determined by the needs of creating a mass market, whereas in Europe they are pitched at the 'executive' class buyer (Table 2.6).

Manufacturers may argue that the new car of 1996 is far better equipped than that of previous years, so that customers are getting greater value than ever before. However, it has to be questioned whether such features as electric seat adjustment, catalytic converters or even ABS are really useful to the average new-car buyer. Furthermore, the extra value has to be paid for, even if not in direct proportion. Some customers cannot afford this, or do not wish to pay the higher actual price.

In the near future we may also see some current car owners abandoning car ownership altogether as owning a car becomes increasingly expensive and restrictions on car use begin to mount up. We already see today that many city dwellers forego car ownership, while others opt for car-sharing schemes (see Chapter 9). This is likely to become more prevalent and parts of the EU may well follow the lead set by certain Asian countries in this respect. This 'high hassle' factor, combined with an excellent public transport infrastructure, is one of the main reasons for the relatively low car-ownership levels in Japan, especially in its urban areas. Of course, market saturation is also a function of the stock of cars in use (or 'parc' as it is known in the automotive industry), a point we return to below.

2.4 STRUCTURAL MARKET CHANGES: FRAGMENTATION AND MARKET SEGMENTS

Market fragmentation is a further threat to the traditional approach to mass producing cars. Fragmentation can be considered as taking a number of forms.

Table 2.7   Illustrative Examples of Dependence upon Key Models, 1994

Company	Model	EU17 sales (000s)	Total EU17 sales (000s)	% dependence
VW	Golf/Vento	721	1214	59.3
Renault	Clio/Twingo	641	1298	49.3
Fiat	Punto	467	1003	46.5
BMW	3 Series	264	387	68.0
Mercedes	C Class	226	418	54.0

Note: EU17 refers to the 15 Member States of the European Union in 1996, plus Norway and Switzerland. % dependence refers to the proportion of total EU sales accounted for by the leading model.
Source: CAIR.

We have already shown that markets have become more fragmented in that the dominant position of one or two leading 'domestic' producers has been eroded. However, within each spatially defined market there is fragmentation in terms of the product segments of the market. In broad historical terms this entails a shift from the market being dominated by the mainstream small- to medium-sized family saloon type cars (typified by the VW Beetle, for example) to a broader range of size categories and a proliferation of body styles (hatchback, coupé, etc.). This process is a clear threat to many vehicle manufacturers because they rely upon one or two key models for the majority of their sales, as is illustrated in Table 2.7.

Such high dependence upon one or two key models means that model replacement is high risk, especially as market fragmentation has increased over the years. Manufacturers are very wary about replacing a model that provides them with half their sales, if not necessarily half their revenue. This phenomenon also accentuates the problem of high fixed costs in vehicle bodies, for although the key models may benefit from economies of scale, other models in the range may not and could represent a cost which can in some cases be cross-subsidised by the higher volume models. Such a policy will lead to loss of revenue and may not be economically viable in the long term.

Market segmentation in Europe

Competing products tend to be most homogeneous in the medium segments (e.g. Ford Mondeo and GM Vectra), where competition is usually at its strongest, volumes high, product cycles short and thus the risks higher. Within segments, sub-segments can also be defined and targeted. Some full line manufacturers maintain product ranges that mirror these segments precisely. The main segments are shown in Table 2.8.

Table 2.8 uses the French system of segmentation and shows clearly that just three main segments account for over 80% of sales. Alternatively, the H2

Table 2.8   Major Segments in Europe, 1994 Registrations

Segment (Ford example)	Registrations (000s)	% of total market
B1 (Ka)	544	4.6
B2 (Fiesta)	3 199	27.4
M1 (Escort)	3 818	32.6
M2 (Mondeo)	2 602	22.2
H1 (Scorpio)	883	7.5
H2 (Jaguar)	65	0.5
Cabriolet	132	1.1
Coupés (Probe)	213	1.8
All terrain (Maverick)	241	2.0
Total	11 697	100.0

Source: adapted from L'Argus de l'Automobile (1995).

segment, which includes the Audi A8, accounted for only 65 000 sales in all — about 0.5% of the market. Of interest are the coupé and cabriolet segments, both of which have shown significant recent growth, although in both cases the absolute size of the segment is small (a combined share of less than 3%). It is also worth noting that this segmentation system does not identify the multi-purpose vehicle (MPV) — typified in Europe by the Renault Espace — as a distinct segment.

The segment most likely to enjoy relative growth in share in Europe is the B1 or Sub-B segment, currently almost exclusively occupied by Renault's Twingo and Italian brands controlled by Fiat. In 1996, Ford launched their Ka model in Europe precisely for this emerging segment, and other vehicle manufacturers are following with their own models.

Segments do not always cover the whole market; they leave gaps, or 'niches'. These are small corners of the market which are not profitable enough for the big volume producers. However, they may be of interest to specialists. Volume producers may be interested if they can enter the niche by shared componentry with volume models (e.g. Renault Twingo) or co-operation with competitors, e.g. the VW Sharan, SEAT Alhambra and Ford Galaxy are all built at the same plant in Portugal.

Sometimes, the niche turns out to be an unexplored segment and once a product is placed into it demand grows so rapidly that a new segment develops. The MPV market is a good example. This was identified as a niche by Chrysler, which commissioned early versions of both what became the Espace and its own Voyager. The Voyager saved Chrysler in the 1980s, when what was considered a niche vehicle built a whole new segment for itself. The same happened to Renault's Espace, developed and built by Matra. This example, shown in Table 2.9, indicates the advantage a manufacturer can gain by being the first in a niche or segment. This first-comer advantage is clearly illustrated by how the Espace managed to retain its share of the home market.

Table 2.9   France: MPV Registrations, 1992–94

Model	1992	1993	1994	% 1994
Renault Espace	30 049	29 145	30 291	57.8
Chrysler Voyager	7 145	7 144	7 560	14.4
Citroën Evasion	—	—	2 582	4.9
Peugeot 806	—	—	5 206	9.9
Ford Aerostar	36	1	—	—
Mitsubishi Spacewagon	621	774	549	1.0
Nissan Prairie	624	198	100	0.2
Nissan Serena	—	3 305	2 982	5.7
Pontiac TransSport	769	1 825	1 316	2.5
Toyota Previa	1 256	627	350	0.7
VW Caravelle	1 809	1 424	1 462	2.8

Source: Taragnat (1995).

The past few years have seen a proliferation of new models, many covering segments or niches that did not exist or were not explored by their manufacturers in the past. As a result, overall volumes produced for each model have been reduced. This is in turn creating a situation where conventional steel body technology, geared as it is to high-volume mass production, is appropriate for fewer and fewer models. Although the process is gradual, it will result over the next decade or so in an increasing number of models for which the alternative body technologies under consideration in this book are more appropriate than the conventional steel monocoque.

This scenario would involve a limited number of standard and basic model platforms and powertrain configurations which could be used to make a larger number of model variants that would be visually distinct to the consumer. In addition, completely new niches will be created by the promotion of alternative technologies such as electric vehicles (see Chapter 5). This sort of development will bring completely new players into the car industry (see Chapter 9).

Market segmentation in Japan

The major sales segments in Japan are determined by the nature of the vehicle classification and taxation regime. This defines cars in terms of overall dimensions and engine capacity, as shown in Table 2.10.

An equally important element shaping car design and sales in Japan is the nature of the taxation regime, which combines national and local elements. Of these elements, variations in local car ownership tax according to engine size are the most important, as shown in Table 2.11.

Japan also has a weight-related 'Tonnage Tax' which is levied on new car purchases at the rate of ¥6300 per 0.5 tonne (with a set rate of ¥4400 for the Kei class cars). In recent years the relative share of the three major segments has

Table 2.10 Car Classification in Japan

	Standard	Small	Mini ('Kei')
Height (m)	>2.0	<2.0	<2.0
Length (m)	>4.7	3.3–4.7	<3.3
Width (m)	>1.7	1.4–1.7	<1.4
Engine size (cc)	>2000	661–2000	<660

Source: JAMA (1995).

Table 2.11 Local Annual Ownership Taxation in Japan, 1995

Engine size (cc)	Annual tax (Yen)
Under 661 (kei)	7 200
661–1000	29 500
1001–1500	34 500
1501–2000	39 500
2001–2500	45 000
2501–3000	51 000
3001–3500	58 000
3501–4000	66 500
4001–4500	76 500
4501–6000	88 000
Over 6001	111 000

Source: JAMA (1995).

changed significantly, as is shown in Table 2.12 (where the growing share of the Mini or Kei class of cars from 1987 is of particular interest).

The rapid growth in the share of the Mini or Kei segment in Japan was in no small way attributable to urban planning and fiscal policies, and the rapid rise in the cost of land in much of urban Japan. This led, for example, to rules limiting car ownership to those who could prove they had a parking space — with Kei cars exempt from the rule. What is interesting about this growth is how quickly it developed in the light of these policy changes, showing that even quite large changes in the structure of the market can be brought about rapidly, given the appropriate economic and policy incentives.

Market segmentation in North America

The traditional segmentation of sub-compact, compact and full size in North America is increasingly inappropriate for current market conditions. The traditional full-size car segment (now classed by *Automotive News* as 'upper mid-range') has declined markedly, causing two major platforms in this segment

Table 2.12   Changing Market Share for Mini, Small and Standard Cars in Japan, 1987–94 (%), Including Imports

Year	Small	Standard	Mini
1987	92.7	3.4	3.9
1988	91.4	4.5	4.1
1989	84.8	6.3	8.9
1990	75.2	9.2	15.6
1991	69.1	13.6	17.3
1992	66.6	16.0	17.4
1993	65.3	16.3	18.4
1994	64.4	16.3	19.3

Source: JAMA (1995).

Table 2.13   US Market Sales by Segment, 1995 (units)

Segment	Best seller	Segment sales 1995
Budget	Plymouth Neon	712 683
Small	Honda Civic	1 743 537
Lower mid-range	Pontiac Grand Am	1 064 667
Mid-range	Ford Taurus	2 638 533
Upper mid-range	Buick LeSabre	774 264
Near luxury	Toyota Avalon	523 561
Luxury	Cadillac DeVille	576 268
Sporty	Ford Mustang	542 616
Specialty	Chevrolet Corvette	57 753
Minivans	Dodge Caravan	1 246 459
Full-sized vans	Ford Econoline	395 810
Compact SUV	Ford Explorer	1 465 880
Full-size SUV	Chevrolet Tahoe	287 535
Compact pick-ups	Ford Ranger	1 011 858
Full-sized pick-ups	Ford F-series	1 722 864

Source: Automotive News Data Center.

— the Chevrolet Caprice/Buick Roadmaster and the Oldsmobile 98/Cadillac Fleetwood — to be phased out without replacement in 1996. In practice, much of the traditional demand for this kind of vehicle has been transferred to light trucks such as 4×4 sport utilities, pick-ups and minivans (see below).

In this context, the leading US car magazine *Road & Track* (1995) has adopted a price-based segmentation system along the lines 'economy', 'mid-priced' and 'luxury', in addition to 'sporting cars', 'sport utilities' (SUVs), 'minivans', etc. The leading trade publication *Automotive News* uses a similar segmentation system based on a combination of size and price (see Table 2.13). The most remarkable phenomenon in the US market over the past two decades has been the way in which downsizing in the car segments has been implemented, making the

Honda Accord, a compact by traditional US standards, the best-selling car in recent years (although for 1995 it was overtaken by the Ford Taurus).

On the other hand, many traditional full-size buyers have transferred to light trucks, thus retaining their ability to enjoy a large V8, V10 or, increasingly, a V6. As a result, the best-selling vehicle in the US these days is the Ford F-series light truck, of which more than half a million units are sold each year. In 1995, 691 452 Ford F series were sold, compared to 366 266 units of the best-selling car, the Ford Taurus. In fact, the best-selling car was only fourth in the ranking of best-selling vehicles in the US in 1995, behind the F-series, the Chevrolet C/K pick-up and the Ford Explorer compact sport utility vehicle (SUV).

2.5   THE CAR PARC AND ITS IMPACT ON NEW-CAR DEMAND

There is a tendency to think of new-car demand as a distinct market, and to ignore the used-car market. However, it is impossible to ignore the impact of the steady expansion in the stock of cars in circulation, the so-called 'parc'. As cars in general become more reliable and last longer, so new cars become relatively less attractive (see Nieuwenhuis & Wells, 1994: chapter 10). While there is a segment of the population that demands new cars, the real demand is for personal *transportation*, or car ownership, that can increasingly be fulfilled with used cars. Thus there is a gradual transformation of the role of the new car, from functional transport provider to fashion accessory; a feature that is likely to make demand for individual models much more volatile.

Table 2.14 shows the number of vehicles in use in selected years, where huge growth is evident. By the year 2000 there will be over 600 million vehicles in the total parc. This global parc is, of course, unevenly distributed: the UK actually has a bigger parc than all of Africa. Table 2.15 illustrates the growth in the stock of cars in selected countries over time, where the mature industrial economies tend to show a decline in the growth rate of the parc in the period 1990–94.

Table 2.14   World Total Registrations (Vehicles in Use) Cars and Commercial Vehicles, 1930–90 (millions)

Year	Cars	Commercial vehicles	Total
1930	29.9	5.9	35.9
1940	37.1	8.9	46.1
1950	53.0	17.3	70.3
1960	98.3	28.5	126.8
1970	193.4	52.8	296.3
1980	320.3	90.5	410.9
1990	444.8	138.0	582.9

Note: totals may not sum due to rounding.
Source: adapted from AAMA (1995).

Table 2.15   Total Registrations (Cars Only) for Selected Countries, 1970–94 (000s)

Country	1970	1975	1980	1985	1990	1994
China	na	na	350	794	1 621	3 000[a]
India	549	673	927	1 338	2 300[a]	3 300[a]
Indonesia	na	na	na	na	1 293	1 870
Japan	8 778	17 236	23 659	27 844	34 924	42 678
South Korea	60	84	249	556	2 074	5 148
France	11 860	15 180	18 440	20 800	23 010	24 900
Germany (W)	13 298	16 763	21 454	23 776	27 212	28 694
Italy	10 181	15 059	17 686	22 494	27 425	29 800
UK	11 801	14 060	15 437	18 952	22 527	23 831
Argentina	1 418	2 310	3 112	3 878	4 352	4 426
Mexico	1 233	2 400	4 254	5 282	6 819	8 500
US	89 243	106 705	121 600	131 864	143 549	147 171

[a]Estimate.
Source: adapted from AAMA (1996).

One consequence of saturated markets and the growth in the stock of cars in circulation is that the inflexibility of the existing vehicle manufacturers' production facilities is exposed. First, there is the emergence of a chronic over-capacity. There is a far greater potential to produce vehicles around the world than there is a capacity to consume vehicles, and yet the process of competition is forcing the vehicle manufacturers to install new capacity (especially in emerging markets) faster than the markets can grow to absorb the available supply (see Pugliese, 1996). The vehicle manufacturers, faced with the high fixed costs of not producing cars once the investments required have been made, are repeatedly tempted to over-produce and hence over-supply the market. This problem was evident in the US market in the late 1980s, and in European markets in the early 1990s. In both cases, the vehicle manufacturers sought ways to 'push' cars into the market, so reducing the pressure being generated by the build-up of stocks of finished cars. In particular, the vehicle manufacturers sought to sell cars into 'fast-turn' segments of the market. In these fast-turn segments, the first user of the car holds it for less than 12 months before it is released for re-sale. Such cars became known as 'program cars' in the North American market. There are three main fast-turn segments:

- the vehicle manufacturers themselves
- the franchised dealership network
- the daily rental industry

The reasoning for this policy was simple. If existing high-volume buyers of cars reduced their average holding periods (i.e. the time between buying a new car and subsequently disposing of it again) then more cars would be sold. Very large discounts (of up to 40% below list price) were available to high-volume buyers, who could then actually re-sell the car for more than they paid for it in the first place (for a UK analysis of this issue, see Wells et al., 1997).

It is not surprising that this practice did not constitute a viable long-term strategy. Large numbers of so-called 'nearly new' cars returned to the market, so depressing further purchases of new cars (i.e. through substitution effects). Equally important, these returning cars were outside the control of the vehicle manufacturers and their franchised dealership networks. This meant that the nearly-new market supported the growth of independent used-car superstores who bought and sold cars as a commodity, concentrating on high turnover and low margins. The net result of this was to drive down the residual values of cars. In the UK case, one estimate put nearly 30% of all new car registrations in 1995 down to fast-turn segments (Wells et al., 1997).

The growing stock of cars shown in Table 2.14 leads to a further point: that new cars become less and less valued and hence less valuable, as expressed in high depreciation rates on new cars and low prices of older used cars. However, at the same time, less and less benefit (or use value) is to be derived by consumers from cars, despite the necessity of ownership. This is due, in part at least, to the sheer numbers of vehicles on the roads, leading to congestion — one of the main causes of the car's environmental problems. There appears to be a transformation under way from consumers actively desiring new car ownership, to one of ownership under sufferance in the face of myriad difficulties; in other words, a major shift in the culture of car consumption which could have profound consequences for the vehicle manufacturers.

While less quantifiable, there appears to be a growth in anti-car social attitudes. This can be illustrated by recent developments in the UK. First there was the horror and outrage which met — admittedly unrealistic — Department of Transport projections for the vehicle population and the associated road-building programme. Then the Royal Commission on Environmental Pollution (1994) report highlighted the social cost of car use. The battles over key new road projects such as Twyford Down and the Newbury bypass have served to highlight the conflict, while the 'Reclaim the Streets' movement has grown in leaps and bounds. This movement engages in direct action by closing off a major urban thoroughfare without notice and organising an impromptu street party, while erecting structures that are obstructive and hard to remove. The movement enjoys considerable public support in many of the city areas it targets (Penman, 1995).

In terms of popular culture, Ben Elton's (1991) *Gridlock* and Heathcote Williams' *Autogeddon* (1991) reflect the view that cars are a bad thing, while academic research has shown that car dependency generates isolation and alienation among young children (Hillman et al., 1990) — only the poor and the vulnerable walk the streets. Even if some of these fears are overdone, the motor industry must respond to these changes in the 'culture' of car consumption and use.

2.6   CRISIS IN DISTRIBUTION AND MARKETING

Consumers as employees perceive themselves increasingly as working in a less secure environment. In addition, they will face higher costs to sustain their standard of living and hence will be more careful with their money. Car buyers will become more sensitive to value for money. As a result, continued growth in consumer power can be expected. Expectations from consumers in terms of the quality of product and service will grow, and there will be less loyalty to brand, especially as brand differentiation is increasingly difficult to justify. Consumers will be more difficult to please, more prepared to complain or take their custom elsewhere. This trend can already be seen in such moves as the formation of a separate Directorate for Consumer Affairs in the European Commission, DG XXIV, the entry of JD Power into Europe, car magazines which are increasingly critical of the industry and even the rising influence of consumer television programmes.

If the character and volume of demand is changing, then existing approaches to the supply and distribution of new cars must also change (Rhys et al., 1995). Traditionally, optimisation of the production system has dominated the picture. Cars were produced whether or not there was demand, because an idle plant costs money. In practice, the unnecessary stock thus created was itself a considerable cost burden, while dealer-swaps (dealers exchanging cars they cannot sell for ones they can, with other dealers) and the discounts offered to induce customers to buy the surplus vehicles further reduced margins and profitability. The high levels of advertising and the prevalence of 'incentives' to consumers to encourage the purchase of new cars is indicative of over-supply.

There are already many initiatives under way at manufacturer and dealership level to adjust to market conditions as they evolve — most notably in the form of 'lean distribution'. In terms of production technology, the actual limits of flexibility in production mean that true demand-led production cannot be used: the factory cannot be turned on and off at will without incurring an enormous cost penalty, and neither are production volumes infinitely elastic. Nevertheless, manufacturers are beginning to try and take leanness even further.

First of all there is the general reduction in customer lead-times. Ford announced as part of its Ford 2000 programme the formation of a worldwide taskforce responsible for cutting vehicle customer-order-to-delivery times to 15 days. The programme is to be piloted in North America, where customer lead-times are typically around 72 days. Volvo is taking a slightly different approach in that it is offering a discount on all vehicles sold as customer orders rather than from stock. This reversal of the traditional approach is also to be pioneered in North America.

Combining lead-time reduction with an increase in true customer orders while reducing stock is not an easy task. Two types of response are evident so far: pipeline and customisation. These are not mutually exclusive. The 'pipeline', as developed by Rover and Vauxhall in the UK for example, means a dealer can order into a moving stock of vehicles, making their way through the production process. It was introduced essentially to improve customer satisfaction and reduce stock, while it also reduces expensive transfer of stock between franchised dealers. Such systems usually work with limited sets of option packs rather than full customisation. In other words, reducing the number of options is being seen as a key part of achieving faster delivery times — but the danger then is that the exact combination of features which an individual customer demands will not be available. Most customers can be satisfied within this system, if it is developed properly. A number of variants on this system exist and several manufacturers and importers in the US and some European countries now operate a central stock of cars from which the dealer may order in a similar way. This then becomes part of the pipeline, although any order of a specification not held in the central stock will still take longer to deliver.

True customisation, such as practised by BMW and other German makers in Germany, trades off delivery time against giving the customer a wider choice of options. Research by the International Car Distribution Programme (ICDP, John Kiff, pers. comm.) indicates that, on the whole, German customers are prepared to wait much longer for the exact vehicle of their choice than other Europeans. However, this high level of customisation carries with it a considerable cost in terms of production complexity. Standardised specification packages are much easier to accommodate and in practice will satisfy most customers. It is significant that after such standardised packages as the 'Reflex', 'Avantage' or 'Volcane' on the Citroën ZX for example, even Mercedes-Benz has introduced specification packages, such as 'Elegance', on its models. Although some choices are possible within packages, they are more limited, carrying less of a production cost penalty.

Alternatives to the franchised dealership network?

Another question is the extent to which selling cars is of interest to established retailers in other areas. In the past, the US department store and mail order house Sears Roebuck has sold cars, while Rovers were sold for a while in the 1980s via the Massa supermarket chain in Germany. However, dealer margins have been squeezed to such an extent that this route may offer few attractions to these retailers today. Firms such as Marks & Spencer would expect a return on sales of around 7%, rather than the 1–2% more commonly found in car retailing. With increasingly cost-conscious consumers on the one hand and manufacturers trying to reduce costs in the network on the other, the prospects for making money are limited, unless a new low-cost formula can be found. The response from the established retail side in the shape of the franchised dealerships is varied. Among the strategies considered are two main strands:

● a large consolidated but specialised site, e.g. a pure sales-only superstore or automall;

- a large consolidated but mixed site, integrating all existing main dealer activities and more.

In any case, site use intensification to achieve turnover increases, and increasing professionalisation of dealer practices, are under way and essential for the future. Much of this is encouraged or forced upon the dealers by manufacturers. Putting all these factors together, we can see significant changes within dealer networks over the next few years, all aimed primarily at reducing costs throughout the system, while retaining or enhancing customer care. The future will see fewer but larger main dealers incorporating all activities; however, these will oversee a network of smaller, more dedicated sites specialising in servicing (including fast-fit), used cars, and bodywork. Some existing dealers will be downgraded in the process.

On the other hand, there is also the danger of new entrants, with new retail concepts and as a result a widening of the competitor base, as in the USA. This will develop first in the more lucrative used-car business, where CarMax in the US is already taking a significant share of the market. If such firms move into new cars (CarMax now has a Chrysler franchise on one site) there is the possibility of comprehensive multi-franchising with several different brands sold on one site. This could lead to price-based off-the-shelf comparison retailing.

All these developments would require a major culture change in an industry still used to forgetting about the product once it leaves the factory gates. The potential market is much bigger. Most people buy used, not new cars and the franchised dealers often lose both used-car sales and service/repair to the independent sector after only a few years. There is a considerable value stream to be tapped into here. Increasing car longevity, already evident and mainly due to the quality improvements of the past 15 years, make this not only more interesting but also more essential as this could ultimately lead to a long-term fall in demand, as it has done in Sweden since the mid-1960s (see Nieuwenhuis, 1994c).

One way in which the vehicle manufacturers and their franchised dealership networks have sought to extend their control further into the life-cycle of their cars has been to create new forms of ownership and payment. These include various fixed term (usually 24–36 months) contract hire or lease packages; in many cases the vehicle manufacturers or their finance division retains ownership of the car. This is an important shift because, alongside the development of 're-marketing' strategies (see Wells et al., 1997), it places the vehicle manufacturers in the position of having a direct financial interest in the running costs, depreciation and longevity of the cars they produce.

Alongside these new personal 'ownership' concepts, there has been a proliferation of experiments in more collective ownership of a reduced number of cars. The collective approach is exemplified by the personal public transport systems using electric vehicles currently on trial by PSA in some French cities and soon to be introduced in Coventry. A similar system was tried in Amsterdam in the 1970s (see Chapter 5). In Edinburgh, a street-based scheme was launched in 1996, whereby all the residents have abandoned car ownership: their mobility needs are met by a small pool of rental cars (see Chapter 9). The collective approach is interesting because a new car makes a poor personal investment: it is not in use about 95% of the time, and yet loses value at the rate of 35–45% per annum. Collective ownership increases the utilisation rate and spreads the burden of depreciation.

Product positioning, price and residual values

In launching a new product, vehicle manufacturers need to have regard for

- their current range and price strategy
- competitor prices
- market segmentation
- extent of product innovation

Many manufacturers consider that consumers have little interest in the material from which their cars are built. There is, then, little or no added value in making occasional hang-on components of the body from materials other than steel — at least as far as the consumer is concerned. In contrast, innovations in drivetrain technology (e.g. CVT, multi-valve engines) or in other parts of the vehicle (e.g. ABS, climate control) offer more tangible and realisable customer benefits or features which a manufacturer can sell. A further interesting point about product features is that their 'value' changes over time. As a feature such as ABS is introduced, it is usually applied to a selected model or variant, usually towards the upper end of the manfuacturers' range. Over time it becomes an option on more of the model range, while it becomes standard fit on the upper end of the range. Eventually it is incorporated as the mainstream technology as a standard fit on all cars in the model range. Initially, when sold as an option on premium cars, manufacturers can charge substantial prices for an 'esoteric' piece of technology. As the product becomes more commonplace, and as production volumes increase, costs fall but so do prices. Moreover, the technology loses its distinctiveness, its ability to differentiate the car and hence command premium prices, and thus features less in marketing and advertising activities.

In the mid-1990s, Audi derived considerable media coverage for their aluminium A8, precisely because it was rather different from all the other vehicles in the segment. Just as important, though less quantifiable, will be the 'halo effect' that spills from the A8 to other cars within the Audi range (which, significantly, have been redesignated the A4 and the A6) and enhances the Audi image of technological innovation. Certainly, both engineers and motoring journalists have been impressed with the A8, and thereby given Audi an important marketing success. However, what is less clear is how durable the marketing advantage will be, although Audi will follow up with another aluminium vehicle, the A1/A2 '3 litre car' (i.e. a fuel consumption of 3 l/100 km).

Moreover, the price that may be charged for a new vehicle depends upon the brand image of the manufacturer and on the residual values that may be expected. Typically, highly specified vehicles with a small market share suffer very high rates of depreciation. In a sense the aftermarket value of a vehicle is the true measure of its worth, and reflects a complex mixture of the functional attributes of a vehicle, its initial cost, the market share of the vehicle, the cost of supporting the car (parts availability, independent repair availability, etc.) and the perceived long-term performance of the manufacturers' product (i.e. product longevity). That is, every manufacturer has a brand image, which tends to vary across different national markets, and which is built up over a long period of time. Recent entry manufacturers (such as Malaysian Proton, or the Koreans) tend to suffer from 'lack of image'. The perceived brand image can seriously circumscribe the options available to a manufacturer when considering innovative product introductions, including the use of new or alternative materials in the body structure. These problems are well illustrated in the case of Mercedes, which is attempting to enter smaller size categories than have hitherto been occupied by the company (i.e. the A Class, the MCC/Smart car).

Brand image is a key concern of vehicle manufacturers, but it is not entirely within their control. Nor does it necessarily respond quickly as a result of marketing and advertising actions. This may militate against the introduction of new body materials, because the manufacturer may not be able to capture the added value with product positioning. It is no surprise, therefore, that novel materials and body designs tend to come from outside the mainstream models of volume manufacturers, who are unable to benefit from such introductions.

2.7   CONCLUSIONS

The combination of saturated and fragmenting traditional markets, inadequate growth in emerging markets, and an investment 'race' to install new capacity have combined to create very unfavourable conditions for mass production. At the same time, the vehicle manufacturers are moving away from a narrow focus on new vehicle sales, towards a longer-term view of product lifetime earnings streams and new ownership concepts are emerging. In combination, these trends are a threat to the traditional approach to manufacturing cars in general, and to the approach to manufacturing car bodies in particular. That is, there is a competitive premium in manufacturing processes which can be more flexible, and profitable at lower volumes, and in products which offer lower lifetime costs. This is the market context which is generating the pressure for a 'paradigm shift' in the automotive industry.

3 The Role of Government

3.1 INTRODUCTION

The automotive industry is no stranger to close government attention, up to and including outright state ownership which in some cases persists. In this chapter attention is drawn to two mechanisms by which government intervention is contributing to the emerging shape of the automotive industry: environmental regulation; and co-funded collaborative R&D programmes. In neither case can the automotive industry be said to be simply a passive recipient of state intervention; on the contrary, policy measures are arrived at through close consultation. Generally, the automotive industry has resisted government attempts to impose product performance standards, with greater or lesser degrees of success. The California Air Resources Board (CARB) initiative which mandated a proportion of sales being composed of zero emissions vehicles has been postponed under industry pressure, as we discuss in more detail below. In evaluating the likely future impact of legislative proposals and the ability of government to impose radical changes on the industry, it is necessary to have regard to the economic and political power the automotive industry can deploy in defence of its interests. Equally, the extent to which 'high profile' co-funded and collaborative R&D programmes can achieve radical technological change might be questioned.

This chapter charts some important trends in the character and scope of government intervention which place further pressure on the established paradigm, while simultaneously offering at least some prospect of innovation. First, the history of emissions legislation is explained. From tentative beginnings, emissions legislation has become both more stringent and wider in scope, culminating in the EU proposals discussed below. More importantly, there is a renewed interest in fuel economy legislation for both environmental and strategic reasons (CEC, 1992; DGXVII, 1996). This sort of legislation will have much deeper 'systemic' effects on the automotive industry. Second, several important R&D programmes have been introduced in the 1990s which seek to bring together industry, academia and government into collaborative projects. In a sense, these programmes represent the ultimate attempt at a 'technofix' to the problems which beset the industry. All may be criticised on a range of grounds. However, these major R&D programmes are probably the best 'signpost' for the technological direction of the automotive industry.

3.2   THE HISTORY OF EMISSIONS LEGISLATION

The history of vehicle emissions legislation starts more from a concern about human health than about the survival of the planet, and like much subsequent legislation it started in California. The concern about the impact of motorised vehicles on our environment centres around four general issues:

- emissions
- energy consumption
- noise
- congestion and land use

All of these have been subject to regulatory and legislative control somewhere over the past few decades; however, most of the debate and regulation has focused on emissions. These emissions from a totally unregulated car are as follows:

- *Exhaust emissions*: these constitute between 65% and 70% of the total. Of these, around 97–98% consists of substances not directly harmful to human health, namely water, oxygen, nitrogen and carbon dioxide. The remaining 2–3% consists of the harmful toxins regulated by law:
  — Carbon monoxide (CO): a product of incomplete combustion.
  — Oxides of nitrogen (NO_x), i.e. nitrogen dioxide (NO_2), nitric oxide (NO) and to a lesser extent nitrous oxide (N_2O): these occur during any high-temperature combustion process involving air.
  — Hydrocarbons (HC): essentially unburnt petrol. They comprise a very large range of chemicals of varying toxicity. Probably most harmful are the aldehydes such as toluene and benzene. Some new chemicals are also formed in the process.
  — Particulate matter (PM): this is mainly a problem of diesel engines. It consists of soot, unburnt lubricating oil, some unburnt fuel, sulphates and wear debris. Petrol engine particulates from engines running on leaded fuel also contain some inorganic lead compounds.
  — Sulphur dioxide and sulphur trioxide are also produced in small quantities by diesel engines.

- *Evaporative emissions*: these are caused by chemicals in the fuel system evaporating into the atmosphere due to leaks or the 'open' nature of vehicle fuel systems. Petrol is particularly susceptible to this.
- *Crankcase emissions*: these consist mainly of lubricating oil mist, combined with some unburnt fuel from piston blowby. The high oil pressure in the crankcase forces these emissions past piston rings and gaskets into the atmosphere.

The primary motivation for government regulation has been that most of these substances are harmful to us, rather than 'the environment'. However, their volume is small as a proportion of total car emissions. One car driving along a country road has a negligible effect on air quality; the problem lies in the numbers of cars and trucks and especially their high concentrations under unfavourable conditions of use in densely populated areas.

As a result, in Europe it is estimated that motor vehicles are responsible for 90% of human-made emissions of CO, 30–40% of NO_x and 40% of HC (Tims, 1990). The figures for the UK alone are 85% of CO, 45% of NO_x, and 28% of HC (Nieuwenhuis et al., 1992).

It was the growth in the volume of traffic that caused the problems, which were first noticed in the late 1940s in the Los Angeles Basin, where peculiar conditions tend to exacerbate the problem. Much of the pollution was attributed to crankcase blowby, and from 1961 the manufacturers voluntarily fitted positive crankcase ventilation (PCV) systems to their new cars sold in California. The State of California made this compulsory from the 1964 model year, thus enacting the world's first vehicle emissions control legislation. At the same time the Federal Government's Surgeon General compiled a report of all known information on the dangers of vehicle emissions, which was presented to Congress in 1963. This led very quickly to the Clean Air Act. This in turn led to the Clean Air Act Amendment of 1965, which enabled the Federal Department of Health, Education and Welfare to set and enforce national standards limiting gaseous emissions from vehicles (Table 3.1).

Table 3.1   The History of Emissions Legislation in the USA from 1963

Date	Action
1963	State of California enacts compulsory fitment of PCV systems Federal Clean Air Act
1965	Clean Air Act amended to enable setting and enforcement of standards
1966	Federal standards adopted for 1968 model year, covering crankcase emissions and exhaust emissions of CO and HC
1970	Clean Air Act Amendments: setting up of EPA, which sets national standards for CO, HC, NO_x, PM and SO_2 in ambient air and at source
1975	EPA deadline for first round of emissions limits, later delayed until 1978 and subsequently delayed until 1980, with some limits not implemented until 1981, hence ...
1981	Final introduction of Federal emissions standards for cars and first particulate limits for truck diesels
1983	Baseline for truck emissions standards
1987	Deadline for limits on HC and CO for trucks, which combined with the car standards are now known as US87
1991	Truck emissions standards tightened up further
1994	Car and truck emissions standards tightened up gradually between now and 2004 Californian TLEV, LEV, ULEV and ZEV, legislation planned to come into force by 2003

Sources: Tims (1990), Nieuwenhuis et al. (1992), CAIR.

Over time, these US standards in turn influenced standards in several other countries. Some countries adopted them without modification, while other countries, such as Japan, have developed their own standards and their own test cycle, although these are still strongly influenced by the US lead.

Europe had been consistently lagging behind the rest of the developed world in terms of emissions legislation. This was partly due to political disagreements within the EC during the 1980s. The introduction of crankcase ventilation systems in 1970/71 and some other minor measures were adopted over time, but the move towards stricter standards, which in the US, South Korea and Japan have been met through the fitting of catalytic converters, was proving controversial, with a split developing on the one hand between the EC and EFTA (with EFTA countries Switzerland, Austria, Sweden, Norway and Finland adopting US87) and on the other hand within the EC.

The split within the EC saw the more environmentally aware member states, notably Denmark, Netherlands, Germany and Greece, favouring stricter standards based on US87; while the UK, France, Italy and to a lesser extent the other member countries favoured a less radical solution centred around lean-burn technology. The European Commission constructed a complex set of regulations to be introduced between 1988 and 1993, but a final decision on these was delayed several times, prompting several member states to act unilaterally and introduce tougher standards or incentives for cars complying with tougher standards.

In April 1989, the European Parliament supported a move by environment commissioner Ripa de Meana towards a simpler regime of tougher standards based on US87, rendering the existing proposals from the Commission obsolete. These measures were introduced for all cars from 1 January 1993 and effectively require the fitment of regulated three-way catalytic converters. Catalytic converters are damaged by lead additives in petrol, so the rapid introduction of unleaded fuel was promoted throughout the EU via incentives in the form of excise duty concessions.

Similar stepped changes in the emissions permitted occurred in Japan, closely linked to the US standards — not least because the US was and remains a key market for Japanese exports of cars. The standards required as of April 1995 are shown in Table 3.2.

The EU also has strict limits on noise emissions, which primarily affect trucks. Considerable investments have been needed by the truck industry to develop and

Table 3.2   Emissions Standards in Japan, as of April 1995

	CO	HC	NO_x	PM	Smoke
Petrol cars	2.70 g/km	0.39 g/km	0.48 g/km	na	na
Diesel cars	2.70 g/km	0.62 g/km	0.72 g/km	0.34 g/km	40%

Source: JAMA (1995).

adopt the various technical solutions to this problem. Incentives have been offered by the Netherlands and Austria.

3.3 FUEL ECONOMY: INCENTIVES AND DISINCENTIVES

Although in Europe, fuel economy has traditionally been achieved through high fuel excise duty rates, in the US, the government has sought to legislate for improved fuel efficiency. Corporate Average Fuel Economy (CAFE) was introduced in 1978 and it sets a maximum average fuel economy figure for all the cars sold by a manufacturer during a given year. If the manufacturer does not reach this figure, a penalty has to be paid, although it can buy credits from a manufacturer that exceeds the standard. The main flaw seems to be that a manufacturer is penalised for the behaviour of its customers. CAFE limits were tightened up regularly until 1990, but have not changed since then. The penalty has become a way of life for companies such as Jaguar and Rolls-Royce. However, the customer does pay a once-off 'gas-guzzler' tax on buying a car that does not reach the standard.

Table 3.3, showing CAFE performance over the period since 1978, illustrates some fundamental problems. The position of the overall standard to be met has changed only 1.5 miles (= 2.4 km) per US gallon since 1983, and has remained entirely static since 1990. In the time since 1978, US producers have shifted from slightly overperforming against the standard, to slightly underperforming. In contrast, imports started the period with a higher fuel economy performance than

Table 3.3   Corporate Average Fuel Economy, 1978–94 (miles/US gallon)

Year	Standard	Domestic average	Imported average
1978	18.0	18.7	27.3
1979	19.0	19.3	26.1
1980	20.0	22.6	29.6
1981	22.0	24.2	31.5
1982	24.0	25.0	31.1
1983	26.0	24.4	32.4
1984	27.0	25.5	32.0
1985	27.5	26.3	31.5
1986	26.0	26.9	31.6
1987	26.0	27.0	31.2
1988	26.0	27.4	31.5
1989	26.5	27.2	30.8
1990	27.5	26.9	29.8
1991	27.5	27.4	30.0
1992	27.5	27.4	30.0
1993	27.5	27.7	29.5
1994	27.5*	26.3	29.4

Note: * = 8.55 l/100 km
Source: Wards (1994).

that which US cars finished with in 1994. In recent years the CAFE performance of imported cars has fallen towards that of the standard — a reflection of the proportional increase in sales of larger and higher performance cars.

CAFE is a very broad type of regulation which is complex to measure and implement. While initially CAFE targets were made progressively higher, the system appears to have stalled in recent years. It has become a highly political piece of legislation as, over the years, the perception of the strategic threat to the US has declined and as imported cars seemed to benefit most (especially Japanese imports of smaller cars).

One concern for the vehicle manufacturers over CAFE in the US is that the standard for light trucks may be raised. This category includes minivans or people carriers, pick-ups and 4×4 off-road 'sport-utility' vehicles such as the Range Rover and Jeep Cherokee. From 1992 the standard has increased only slightly, from 20.2 miles per gallon to 20.6 miles per gallon (= 11.4 l/100 km). At the same time, the large shift in purchases towards light trucks has continued, which reinforces the attention given to truck CAFE standards.

3.4   THE INCENTIVE APPROACH

European governments have historically favoured incentives, rather than disincentives, in order to bring about what they perceive as more environmental buying behaviour on the part of their citizens. Table 3.4 illustrates European incentives designed to adjust the behaviour of consumers through market signals.

Table 3.4   Environmental Incentives in Europe, 1985–95

Country	Incentives policy applies to
Denmark	cars fitted with catalytic converters, scrappage incentive
Netherlands	'clean' cars, 'clean' and low-noise trucks
Greece	'clean' cars, scrappage incentive
Germany	'clean' cars, catalytic-converter retrofitment, electric vehicles
Switzerland	'clean' cars
Austria	'clean' cars, 'clean' and low-noise trucks
Sweden	'clean' cars
Norway	'clean' cars
Finland	'clean' cars
UK	electric vehicles
Ireland	electric vehicles, scrappage incentive
Portugal	electric vehicles
France	scrappage incentive
Spain	scrappage incentive

Note: the definitions of 'clean' vary from country to country.
Source: adapted from Nieuwenhuis (1994b).

There are few grounds for considering that scrappage incentives are an environmental policy (see Nieuwenhuis & Wells, 1996), although proponents contend that they remove the older, 'gross polluting' cars from the road (SMMT, 1995). Scrappage incentives generally offer owners of old cars a sum of money if they scrap it and buy a new one, and in practice vehicle manufacturers have matched such incentives with their own, giving consumers a further benefit from scrapping old cars. While this may be of value in countries such as Greece and France, which have a relatively old stock of cars in use and a lax or recently introduced vehicle-testing regime, it is of far less value in countries such as the UK where vehicles face an annual emissions performance test. In any case, scrappage incentives fail to take a life-cycle view of vehicle environmental cost — an important consideration when perhaps 25% of total energy consumption is accounted for in vehicle production and final disposal (Nieuwenhuis, 1994b; Wells & Nieuwenhuis, 1996; Wells et al., 1996c).

3.5   CARBON DIOXIDE

Considered from an historical perspective, it is clear that rapid policy formulation may follow only where a clear scientific consensus on the nature of the problem and potential solutions to that problem are available. Additionally, the cause of regulation is greatly aided where an economic cost may be attributed to environmental damage; for example, health care costs may be calculated when considering intervention to improve urban air quality (OTA, 1995). Thus, for example, the Montreal Protocol to phase out the use of CFCs in aerosols, air-conditioning systems and certain production processes (widely used in the electronics and to some extent also automotive industries) was possible because (a) there was a definite link with the depletion of atmospheric ozone levels; (b) such depletion was harmful to human health; and (c) alternatives to CFCs could be employed at relatively little extra cost. A similar case could be made in the instance of asbestos in brakes, clutches and heat shields.

With global warming, however, the scientific consensus is less obvious both in terms of the mechanisms by which CO_2 emissions are linked to climate change and the relative importance of the automotive contribution to that process (see, for example, Metzner, 1995). This allows an automotive industry view, as expressed by the representative organisation ACEA, that

> The proponents of the pessimistic school employ climatic prediction models which suggest a catastrophe. The proponents of the opposing theory, on the other hand, believe that the rise in carbon dioxide concentrations can be counterbalanced by nature itself and therefore do not think that it could have a noticeable effect.
>
> (Diekmann, 1994: 8)

With respect to global warming, in 1996 the Intergovernmental Panel on Climate Change (IPCC) for the first time definitively concluded that not only does global warming exist as a phenomenon, but also that the balance of evidence supports the contention that human agency is directly involved. None the less, it is evident that in most cases there is no precautionary principle in our approach to environmental damage and its consequences. That is, policy only emerges after it has been shown, beyond reasonable scientific doubt, that a dangerous cause-and-effect relationship exists (Rowlands & Greene, 1992). This causes delay, and allows industry to continue to support the status quo. Despite these difficulties, the European Union committed itself at the 1992 Rio Summit to stabilise CO_2 emissions by the year 2000 at 1990 levels, and thereafter to reduce them. The automotive industry itself has produced figures which suggest that in the European Union road transport contributes some 600 million tonnes of CO_2 emissions per annum (Diekmann, 1994: 9). These emissions are produced in the fuel combustion process, but are not directly toxic in the way that other 'controlled' emissions are. In its own research, the European Commission considered that the road transport sector as a whole represents 14% of total CO_2 emissions in Europe and, more ominously, the transport sector is one of the few in Europe showing growth in CO_2 emissions: up 43% between 1980 and 1993 (EAE, 1995). The European Commission estimated that, under a business-as-usual scenario, CO_2 emissions from cars would grow by about 20% by the year 2000, and 36% by 2010, from 1990 levels (European Commission, 1996).

CO_2 emissions from internal combustion engines are directly related to fuel consumption. However, it is also the case that recent regulatory policy innovations demanding the use of side-impact systems and catalytic converters have actually increased vehicle weight and hence fuel consumption and CO_2 emissions.

The principle of the proposed EU energy tax is quite straightforward. Vehicles that have fuel consumption over a specified target would be penalised by an extra purchase tax, which would provide a clear economic incentive for consumers to reduce demand for such vehicles. The proposals are summarised in Table 3.5.

The performance target would have to be met by 2005. It is further proposed that the energy or carbon tax would be sensitive to different size or weight classes of vehicle. Other features of the proposed EU carbon tax worthy of note are as follows:

Table 3.5   The EU Energy Tax: Purchases and In-Use Proposals

Target performance	5 litres/100 km (petrol)
	4.5 litres/100 km (diesel)
Purchase tax	1050 ECU/litre per 100 km over target (petrol)
	1180 ECU/litre per 100 km over target (diesel)
In-use tax	140 ECU/litre per 100 km over target (petrol)
	160 ECU/litre per 100 km over target (diesel)

Source: European Commission (1996).

- A 'no regrets' stance which suggests substantial improvements in fuel economy can be made without additional cost. Consumers would recover the higher initial purchase price through reduced spending on fuel over the lifetime of the vehicle.
- Links to other policy measures, e.g. a reduction in car use.
- Improved industrial competitiveness through energy efficiency.
- Proposed scrappage incentives to remove older vehicles from the stock of cars in circulation.
- A refusal to allow fuel efficiency measures to compromise safety standards.
- A recognition of the need for any proposed policy to be equitable in social terms.

Many of the details of the proposal have yet to be decided. For example, in the discussion document released by the European Commission, no mention was made of the test cycle over which fuel economy would be measured. Clearly, however, within these proposals there is a danger that certain manufacturers will be penalised — notably manufacturers with a strong presence in the larger-car segments. European manufacturers, including Volvo, Mercedes, Porsche, Jaguar, Audi, Saab and BMW, derive a large proportion of total sales from vehicles with engines over two litres. Equally, it is not clear that use of the price mechanism in itself will be sufficient to achieve the aggregate fuel economy targets.

While there may be many 'operational' difficulties in establishing the carbon tax, there are also two fundamental weaknesses. First, the proposal makes no attempt to account for the emissions of CO_2 through the whole life-cycle of the vehicle. New materials, technologies and design approaches may reduce in-use fuel consumption (and hence CO_2 emissions) but actually increase emissions in the manufacturing process. Second, the approach remains partial in that only one issue is tackled, rather than (the admittedly even more difficult issue of) legislating for a more generally environmentally optimised vehicle or EOV (see Chapter 7).

An interesting feature of these proposals is that their effect will be felt throughout the vehicle. This is because vehicle weight is a fundamental factor in determining fuel economy performance. Over time, vehicles have shown a consistent rise in average weight, which has offset continued improvements in engine performance (Nieuwenhuis et al., 1992: 41–3; Nieuwenhuis & Wells, 1993). Thus, to meet fuel economy standards, the automotive industry will have to institute weight reduction as a priority — a major threat to the all-steel body which is a significant contributor to overall vehicle weight. On the average car of 1000 kg, the body plus closures (doors, etc.) can account for 250–300 kg.

An additional factor facing the structure of the market in Japan is the fuel efficiency targets of the government (MITI), announced in 1993. Since the early 1980s, average fuel consumed per vehicle in Japan has remained almost constant, despite improvements in engine performance, because of the growth in sales of large cars. Japan has to import all of its fuel needs for cars, and in 1994 total

Table 3.6 Fuel Efficiency Targets for the Year 2000, Japan

Vehicle weight (kg)	Class	Target (km/l)	Improvement (%)
Under 827.5	Mini and small	19.0	7.3
827.5–1515.5	Small	15.0	8.3
Over 1515.5	Standard	9.1	11.0

Source: JAMA (1995).

imports of fuel (for all uses) amounted to US$47 million or 17% of total imports (JTERC, 1995). The fuel economy targets for the year 2000 are shown in Table 3.6.

3.6   THE PARTNERSHIP FOR A NEW GENERATION OF VEHICLES

> PNGV represents the opportunity to more efficiently address fundamental national objectives than the regulatory mandate approach that we have taken before.
> (Tim Adams, PNGV Director for Chrysler)

Probably the most influential of all government R&D programmes that will have a bearing on the future of automotive technology is the Partnership for a New Generation of Vehicles (PNGV). It represents a change of approach in the USA, from broad regulation (as in CAFE) towards a government–industry partnership to generate new technologies. The overall 'package' of the PNGV is very complex, with many federal governmental departments and agencies being involved, and a range of governmental initiatives have been brought together. In any case, the PNGV shows clearly the main issues that have to be addressed in terms of materials choice, vehicle body structures and powertrain in the broader context of innovations in other related automotive technologies. The PNGV has three central goals:

- to significantly improve national manufacturing competitiveness
- to implement commercially viable innovation from ongoing research on conventional vehicles
- to develop a vehicle that will achieve up to three times the fuel efficiency of today's comparable vehicle

The PNGV has further elaborated the character of the vehicle that the programme should deliver (PNGV, 1994). In addition to achieving 80 US mpg ($=2.9$ l/100 km), it should

- comply with Clean Air Act requirements at the time of production
- meet safety standards of the day
- carry six passengers in comfort
- accelerate from 0–60 mph (0–100 kmh) in 12 seconds

- carry 200 US lbs (90 kg) of luggage plus six people
- have a metro highway range of 380 miles (611 km)
- achieve 80% recyclability

The PNGV has identified a range of 'technology areas' within which innovations are needed, as well as some of the candidate technologies:

(a) Advanced lightweight materials and structures where candidate technologies could include:

— design optimisation
— high-strength steel
— polymer matrix composites
— metal matrix composites
— ceramics
— engineering plastics
— aluminium, titanium, magnesium
— joining technologies and adhesives
— recycling
— process/cycle time advancements in manufacturing

(b) Advanced manufacturing, including:

— agile manufacturing
— high-speed data communication and management
— rapid prototyping and virtual manufacturing
— high-performance computing
— supercomputing
— advanced forming technologies
— advanced joining technologies

(c) Powertrain systems, both in terms of more efficient internal combustion engines, and alternative powertrain systems:

— lean-burn internal combustion engines
— fuel cells
— hybrid propulsion systems
— high-power batteries
— flywheels
— ultracapacitors

The PNGV has identified an approximate design space within which trade-offs between reduced mass and improved thermal efficiency in engines would deliver a vehicle which meets the PNGV performance targets. In simple terms, mass reduction can be sacrificed if powertrain improvements yield over 50% thermal efficiency. Alternatively, small gains in powertrain thermal efficiency will demand much greater reductions in vehicle body weight. According to the PNGV Secretariat:

Even with the improved power converters and regenerative braking, reductions in vehicle mass on the order of 20% to 40% from today's baseline vehicles are required. These levels of mass reduction are beyond simple refinement of today's steel frame, steel body construction and will involve the introduction of entirely new classes of structural materials. (PNGV, 1995: 8)

This indicates the extent to which the PNGV could influence the future of material choice for vehicle body structures and automotive technology generally. The part of the PNGV concerned with advanced materials is of direct relevance to the future of steel, aluminium and plastic composite in vehicle body structures for example, while that concerned with powertrain is seeking alternatives to the internal combustion engine. The PNGV intends to make progress in four interrelated areas:

1. enhanced processing and manufacturing techniques for lightweight materials;
2. similar advances for 'high tech' materials such as metal matrix composites;
3. recycling technology, both in-process and post-consumer;
4. robust analysis and design technologies.

However, any reduction in vehicle mass should not compromise comfort, safety, handling, performance, reliability or cost. The PNGV regards potential replacements of mild steel as being likely to arise from high-strength steel, aluminium, polymer-based composites, metal matrix composites, and magnesium, but each has 'substantial barriers' to overcome. It is worthwhile to review some of the perceived barriers, potential solutions and specific inventions which the PNGV consider important. Table 3.7 shows the main challenges in the areas of materials and advanced manufacturing relevant to materials choice and vehicle body structures.

The PNGV has clear milestones built into it, with a commitment to defined targets within each phase. The first task is to sift the available candidate technologies with a view to concentrating efforts on the most promising technologies. The task of narrowing the technological focus was targeted for completion by 1997. From this point resources will only be committed to those technologies which have been selected. From 1997 to 2000, the aim is to produce a range of concept vehicles to demonstrate initial product viability at component level.

Vehicles produced at this stage are not likely to embody a comprehensive solution; they will be rolling test-beds for a diversity of component technologies. However, in the period 2000–2004 the intention is to prepare cohesive prototypes which will seek to integrate technology concepts into an overall vehicle design capable of production. The intention, then, is to develop collaboration from pre-competitive research into product development — much nearer the market than has hitherto been the case.

Table 3.7 Challenges, Potential Solutions and Inventions Needed in Materials and Advanced Manufacturing for the PNGV

Challenge	Potential solution	Invention needed
Lack of low-cost, high-strength, high-stiffness materials	High-modulus fibre materials	Low-cost carbon-fibre materials
	High-strength, high-stiffness resins	New cost-effective resins
Lack of cost-effective, high-volume production processes for polymer composite parts	Fast-moulding process	Quick-acting catalysts or rapid-curing resin
	High productivity equivalent to hand layering	Automated process to put orientated fibres into place
	Rapid curing of prepreg assembly	Rapid-curing resin or fast-cure method, e.g. X-ray
High cost of metal matrix composites	Develop lower-cost materials	Economical high-volume methods for metal matrix composites
Lack of robust modelling systems in R&D	Development of modelling and simulation methods	Simulation and validation of manufacturing
		Alternative component and product development cost models which account for durability
Lack of inter-operability and flexibility of machines	Open architecture controllers	Controllers which accept modular hardware and software from multiple sources
		Controllers which accept plug and play type sensors
Fixtures too dedicated and labour-intensive	Agile part fixturing	Flexible fixtures for assembly
Joining aluminium components	Suitable mechanical fasteners	Practical rapid clinching or riveting
	Low-energy spot welds	Laser-based spot welds
	Practical adhesive bonding	Adhesive with consistent bonding with little surface preparation

continues overleaf

Table 3.7 (*continued*)

Challenge	Potential solution	Invention needed
		Sensors to detect defective bonds in production
	Technology to assure reliability of spot welds	Control of distortion and part cleanliness
		Equipment to handle variety of joint configurations
		Sensors and feedback to ensure integrity of spot welds
Stamping processes are cost-effective for high-volume parts	Rapid tool and die fabrication	Technologies for rapid tool and die development
		Rapid tool and die fabrication
		Improved predictive modelling of parts size and die design
Specialised CAD tools	Flexible modelling systems	Standard language to allow information exchange across CAD systems
		User interfaces
Excessive cycle times and scrap in aluminium casting		

Source: PNGV (1994).

Structure and membership of the PNGV

The entire PNGV programme is a complex mixture of government agencies and laboratories, universities, and large and small enterprises. In its first phase the PNGV essentially consists of the 'relabelling' of existing research projects. However, as the PNGV develops, so projects will be specifically commissioned towards it.

The detailed operation of the PNGV will not be described here, but the basic outlines will illustrate the scope and nature of the initiative. The main governmental agencies involved are shown in Table 3.8.

Table 3.8   Agency Involvement in the PNGV

Agency	Body involved	Funding programme
Department of Commerce	National Institute of Standards and Technology	Advanced Technology Program
		Energy Related Inventions Program
		Small Business Innovation Research (SBIR)
Department of Defense	Advanced Research Projects Agency	Technology Reinvestment Program
	US Army Tank Automotive Command	Broad Agency Announcement
		SBIR
		Small Business Technology Transfer
Department of Energy	National laboratories	Energy Related Inventions Program
		Program Research and Development Announcement
		SBIR
		Small Business Technology Transfer
NASA		SBIR
		Small Business Technology Transfer
Office of Management and Budget		
Department of Transportation		SBIR
Department of the Interior		
Environmental Protection Agency		SBIR
National Science Foundation		SBIR
		Small Business Technology Transfer

Source: PNGV (1994).

All of the agencies may use CRADA (Co-operative R&D Agreements). These are 50/50 cost-shared industry–government laboratory partnerships in which the government element is contributed in kind. Membership of any PNGV consortium or project is limited to US companies and research bodies, except in a minority role or as subcontractors.

The governance structure is controlled by an Operational Steering Group, led by the Department of Commerce, with representatives from all the relevant governmental agencies, senior management staff from GM, Ford and Chrysler, and USCAR. USCAR is a collaborative forum for Ford, GM and Chrysler. This steering group determines strategy and takes all key technology decisions. The steering group is supported by a technical team, whose role is to provide more detailed management and evaluation of the PNGV. Finally, there is a government PNGV secretariat, acting as an information compilation and dissemination body, and USCAR, which administers the industry involvement in the PNGV (PNGV, 1995).

Hitherto, R&D collaboration between the major automotive assemblers in the United States had been constrained by a combination of powerful anti-trust legislation on the one hand, and the strong competitive rivalry between the firms on the other. Through the 1980s the context for GM, Ford and Chrysler changed, as Japanese competition and environmental challenges provided a twin assault on market dominance. Perhaps more importantly, huge changes were sweeping the US military–industrial establishment in the light of real reductions in military spending, leading to defence-aerospace companies and state R&D laboratories looking elsewhere to apply their expertise (Wells, 1996b).

As noted above, USCAR was formed in 1992. However, it was founded around several pre-existing programmes of collaborative research. The main programme areas are shown in Table 3.9.

One of the consortia most advanced in its work is the USABC (US Advanced Battery Consortium). This has already awarded first phase contracts and is moving into the next stage in the development process. First phase contracts were essentially feasibility or 'proof of concept' studies which, subject to results, provide the basis on which second phase contracts are decided. For example, 3M won an initial US$32.9 million contract for first phase research into thin-film lithium polymer batteries, which showed the viability of the concept. The contract ran from 1993 to the end of 1995. In February 1996, the company, won a US$27.4 million contract to construct multiple thin-film batteries into a modular pack which could form the basis of an electric vehicle. In both cases the contract was awarded by USABC, with funding from 3M, USABC, the US Department of Energy, Argonne National Laboratory and the Canadian power utility company Hydro-Quebec. The view expressed by the Chief Executive Officer of 3M is equally illustrative:

> Becoming the first country in the world to develop a marketable electric vehicle would create a significant advantage in the global marketplace. It would help create

new jobs, positively impact the environment and enhance the automotive industry. (L. D. DeSimone, 20 February 1996, on http://www.mmm.com/profile/pressbox/contract.html)

Significant resources are being poured into PNGV efforts. In September 1993, GM launched a US$148 million, five-year cost-shared programme to develop a hybrid-electric vehicle (HEV) and has appointed a host of supplier companies to conduct research into batteries, flywheels, fuels, motors and controllers, Stirling engines and turbine engines (GM, 1996). Ford, in December 1993, launched a US$138 million HEV programme, 50% funded by the Department of Energy, and

Table 3.9 USCAR Consortia

Consortium	Formed	Area of research
Automotive Composites Consortium	1988	Structural polymer composites
Auto/Oil Air Quality Improvement Research Program	1989	Vehicle emissions and ozone pollution
Vehicle Recycling Partnership	1991	Recycling, reuse and disposal of components
Environmental Research Consortium	1991	Evaluation of environmental impact
US Advanced Battery Consortium	1991	Advanced energy systems for electric vehicles
CAD/CAM Partnership	1992	Product/process design and tooling
Low-emissions Technologies R&D Partnership	1992	Emissions control
Occupant Safety Research Partnership	1992	Impact performance analysis
Low Emissions Paint Consortium	1993	VOC emissions from paint processes
US Automotive Materials Partnership	1993	Materials and processing
Supercomputer Automotive Applications Partnership	1993	Computing and processing for automotive R&D
Natural Gas Vehicle Technology Partnership	1994	Natural gas powertrains
Electrical Wiring Components Application Partnership	1994	Standardised connection systems

Note: High Speed Serial Data Communications Partnership was started in 1991 and concluded in 1994.
Source: USCAR (1995).

has similarly awarded contracts to supplier companies in various areas including fuel cells and ultra capacitors (Ford, 1996). In this particular case Chrysler were a bit later with their programme, launched in February 1996 as a cost-shared US$84.8 million programme for a production-ready HEV propulsion system (Chrysler, 1996). HEV developments are only one part of the overall PNGV effort: similar studies are under way with new or improved materials, metal matrix composites, agile manufacturing systems and much more.

3.7 OTHER COLLABORATIVE R&D INITIATIVES: EUROPE AND JAPAN

EUCAR and the EU car of tomorrow

The primary focus of co-operation for the European automotive industry has, historically, been upon issues of economic policy within the EU (including, for example, block exemption) and trade issues — notably with respect to imports of Japanese cars to the EU. These issues tend to remain the main focus of co-operation in the EU but, in parallel with EU growing interests in environmental issues and the use of pre-competitive, collaborative R&D, the European car manufacturers have sought to develop mechanisms and policies for co-operation, ultimately leading to the creation of EUCAR.

> The industry's worldwide competitors (mainly in Japan and the US) are responding by launching large-scale cooperative R&D initiatives with government assistance. It's simple; if European industry is to compete, it must respond with better targeted and closer-to-market research. It has taken the first step by forming the European Council for Automotive Research and Development, EUCAR, as an agent for expanding collaborative effort on basic research.
>
> (ACEA, 1994b: 2)

In 1980 the European automotive industry leaders formed the Joint Research Committee of the European Automotive Manufacturers (JRC) as a limited and non-strategic forum to promote shared work in such fundamental areas as advanced materials, gas flow simulation, durability prediction and many more. The JRC acted in a co-ordination role and gave some support for projects in, for example, project management.

By the early 1990s vehicle manufacturers in Europe saw a need to extend the scope of the JRC in the light of growing competitive and environmental pressures on the industry. While the JRC was seen as a scientific success, what was needed was an approach that would give competitive success for European manufacturers by strategic co-operation of R&D. This gave rise to EUCAR, formed on 27 May 1994. The members are BMW/Rover, Daimler Benz, Fiat, Ford of Europe, Opel, Porsche, PSA Peugeot Citroën, Renault, Volkswagen Group and Volvo.

EUCAR is closely linked to ACEA (the Association of European Automotive Manufacturers); indeed at present the two organisations share the same building.

EUCAR does not undertake R&D itself. As with the USCAR organisation, EUCAR consists of a very small full-time secretariat under the control of a Council, which is itself linked to the Board of Directors of ACEA. The EUCAR secretariat in turn co-ordinates eight thematic groups:

- materials, structures and related processes
- engines, fuels and exhaust treatment
- manufacturing processes and organisation
- electric–hybrid vehicles and components
- advanced vehicle components and systems
- vehicle development methods and processes
- electronics
- road traffic systems and management

Additionally, *ad hoc* groups are established to perform a specific task; for example, there is one on recycling because it cuts across the thematic groups and because it is seen as a matter needing specific attention at the present time. While control is in the hands of the vehicle manufacturers, it is not intended to exclude the supply industry, infrastructure providers or others from the various research projects that EUCAR is developing.

EUCAR has developed an overall R&D masterplan to provide a common basis for co-operation. Its aim is to give a consistent and structured approach towards radical technological improvement, but, unlike USCAR, it does not have specific goals in terms of how vehicles should perform. The overall strategy is shown in Table 3.10.

The EUCAR approach also links in with other pan-European initiatives such as the ACEA Components Standardisation Group formed in 1993, and the European Programme on Emissions, Fuels and Engine Technologies (the Auto-Oil Programme).

The main source of funds for EUCAR-type collaborative research is the European Commission. The Commission launched its '4th Framework' call for research proposals in 1995/6 (the 1st Framework began in 1984). The vehicle manufacturers, during the definition phase of the 3rd Framework, submitted a sectorally focused proposal in the form of the Environmentally Friendly Vehicle Programme. This was rejected by the Commission, but none the less remains the basis of the industry approach to gaining European funding.

It is clearly too early to tell whether EUCAR will make any impact. While there is potential in some of the thematic groups to influence the question of material choice and body structures, no details of specific research projects are available. The close link with ACEA suggests that the main focus may be on providing information for lobbying the European Commission and other governmental organisations in a 'defensive' role.

While the European Commission remains the main source of funds for collaborative EUCAR projects, there are some difficulties with the type of research the Commission supports and the management of frameworks. From an

Table 3.10   EUCAR Overall Strategy

Strategic Objectives
Environmental sustainability
Customer requirements
Economic objectives
Social acceptance

Basic Tasks
Fuel economy
Emissions reduction
Weight reduction
New methods
New technologies
Conservation of resources
Customer-orientated traffic concepts

Approaches
Integrated vehicle technologies
Advanced IC engines
Materials and related technologies
Technologies for control systems
Technologies for traffic management
Advanced electric and hybrid traction
Manufacturing engineering
Future management concepts
Future organisational structures

Source: ACEA (1994b).

industry perspective there are complaints that the 4th Framework is too biased towards academic, basic and non-applied research. The industry is equally critical of bureaucracy, which slowed down the rate of technological progress. Lastly, the industry contends that the scale of the collaborative effort is inadequate compared with the programmes under way in Japan and the US. There is merit in all of these criticisms of the European approach. To date, collaborative R&D in Europe appears too constrained and marginal to make a profound impact on the technological direction of the automotive industry, at least in the short to medium term (Wells, 1996b). Much depends upon whether the automotive industry in Europe can muster the political support for increased sectoral R&D spending.

Although the European Union does not as yet have a comprehensive vision of the future to match that presented by the PNGV, there are moves towards creating that shared vision. To this end, three European Commissioners have started the establishment of a Task Force on the Car of Tomorrow whose aim is 'to facilitate the research and demonstration work required to develop a competitive "car of the future" very quickly (by 2000–2005)' (*Car of Tomorrow Newsletter*, No. 0, 1995: 2).

The Commission has identified the critical technological factors to be addressed for the rapid development of the Car of Tomorrow, including:

> Advanced energy storage and propulsion technologies, particularly those relating to batteries and fuel cells; essential accompanying technologies (electronics, lightweight materials, telematics, etc.); combining these technologies in zero-emission or hybrid vehicles in close cooperation with the representatives of the motor industry.
>
> (*Car of Tomorrow Newsletter*, No. 0, 1995: 2)

The MOSAIC project

In 1990, Renault, in partnership with several European components and materials producers, launched a project for Material Optimization for a Structural Automotive Innovative Concept (MOSAIC), as part of a wider EC initiative known as EUREKA. This project, with over 70 researchers involved and a budget of around FF360 million, reported its results at the end of 1994. As with many collaborative research initiatives, the additional cost of working with many partners was mitigated by the reduction in risk associated with shared development.

MOSAIC was unusual in that it sought to develop two parallel lines of research with respect to vehicle body structures: a near-term refinement of existing steel technology that would yield immediate and realisable results with contemporary production technology; and a longer-term, more radical 'hybrid' body structure using semi-finished aluminium products and subassemblies in composite materials. As such, the MOSAIC project is almost a microcosm of prevailing European thinking on choices for vehicle body structures. More efficient steel monocoque bodies are seen as offering potential for the next generation of vehicles; alternative materials and designs are still not viable for mass applications, but are seen as long-term solutions.

For both elements of the research project the Renault Clio was taken as the main reference vehicle. In the MOSAIC project the focus was on the vehicle structure not including the 'outer shell', and not including 'hang-on' parts (doors, etc.). The main aims of MOSAIC were

- reduction of fuel consumption
- subsequent reduced environmental impact
- recyclability
- model diversity
- acceptable cost

In summary, MOSAIC sought to optimise the weight–performance–cost equation in the vehicle body structure. Table 3.11 summarises the membership of MOSAIC and the main contributions made towards the project. The membership reflects the two elements of the project noted above.

The first route to improving the performance of vehicle body structures in the MOSAIC project sought to develop more efficient use of steel, validated on a

Table 3.11   The MOSAIC Consortium

Company	Country	Main automotive interests	Contribution
Renault	France	Volume producer of passenger cars	Clio reference vehicle, overall project management, key targets, design integration
Ciba Geigy	UK	Production of structural elastomer adhesives	Steel adhesives, composite adhesives
DSM	UK	Thermosets, thermoplastics, elastomers	High modulus compound (HMC) front end assembly
Enichem	Italy	Polymers	Fibreglass-reinforced polyurethane floorplan
Hydro Aluminium	Denmark	Aluminium extrusions	'Bird cage' structure, front structural module
Montedison	France	Composite materials	Materials for floorpan
Sollac	France	Wide strip steels	Soundproofing steels, high-strength steels

Note: country denotes location of main division within the group responsible for MOSAIC work, not necessarily country of origin.
Source: Renault (1995).

Clio-based prototype. Three core technologies were investigated:

● high-strength steels
● anti-vibration sandwich steels
● elastomer bonding

Some use of laser-tailored blanks (i.e. flat sheets of steel made by welding together two or more separate parts with lasers) also featured in the steel MOSAIC prototype, for example on the front chassis rail.

There has already been some growth in the application of high-strength steels in vehicle structures, but thus far only in selected applications. On a typical Clio, high-strength steels represent 8% of total structure weight; in the MOSAIC prototype this was increased to 17%. High-strength steels allow a thinner gauge of steel to be used, thus delivering weight savings. In one specific area, the suspension turret, substitution saved 25% of the original weight (i.e. 800 g). Costs did not increase significantly, though it is worth noting that these steels do present some production difficulties, such as significant 'spring-back', and that they do enjoy a price premium over basic mild steels.

In the automotive industry the use of sandwich or laminated steels has thus far been limited mainly to Japanese cars. Sandwich steels consist of two thin sheets of steel bonded with a layer of resin between them. However, Sollac (the leading supplier of sheet steel in France) has licensed this technology and is looking for European applications, so to this extent the MOSAIC project offered a useful opportunity to demonstrate the material capabilities. The main value of sandwich steel application is that it can reduce or eliminate the need to use heavy non-woven sound insulation materials — materials which can account for 50 kg of vehicle weight. The galvanised sandwich steel was applied to the dashboard panel, central and rear floorpan and rear wheel housings, 18% of the structure weight. However, the material has severe limitations in forming and joining; thus the development of better epoxy adhesives for steel bonding was an important part of realising the material's potential.

Ciba Geigy produced an epoxy adhesive for the assembly of steel sheet which reduced by 25% the number of spot welds required, improved the rigidity and durability of joints, and reduced the amount of mastic used after the welding process to seal joints. For metal-to-metal bonding, both steel and aluminium, Ciba Geigy provided a single-component adhesive which is heat-cured after application, using the Araldite brand name. Renault estimate that the use of structural adhesive bonding improves the rigidity and endurance of steel structures by 10–15%.

The steel optimisation route delivered a 10% reduction in vehicle structure weight compared with the Clio, without compromising the rigidity and NVH (noise, vibration, harshness) levels specified for the production vehicle. Handling, endurance, impact and corrosion performance were all comparable with the existing Clio reference vehicle. Material costs would be higher than standard steels, but this would be offset to some degree by savings in other materials (such as the soundproofing) and by simpler assembly processes. Overall costs would be less than 5% greater than those for the current Clio, assuming a rate of production of 1800 vehicles per day, but there would be no cost advantage associated with lower production volumes as the tooling costs are essentially the same as for the standard Clio. However, in the view of Renault, 'this solution reveals the limits of steel as a material — even though low cost represents progress in the right direction, weight saving limited to 10% does not go nearly far enough' (Renault, 1995: 16).

In the light of this finding, the MOSAIC project also tried a much more radical approach with plastic composites and aluminium. It is interesting to note that Renault did not follow an all-aluminium concept in the MOSAIC project, but sought instead to develop a 'hybrid' or multi-material body structure. Two versions of the front end of the Clio were developed, one in aluminium and the other in composites. As Renault note, 'This approach represents a major technological break with the past, which would create a major upheaval in the traditional design and production processes and would mean that all the functions involved in the development of a new product — from design through to production and after-sales — would have to be redesigned' (Renault, 1995: 18).

In the multi-material, steel-free structure, Renault used a 'bird-cage' upper structure made entirely in aluminium, consisting of strips of extruded aluminium with roof and quarter panels made from aluminium sheet. The dash panel and, most interestingly, the floorpan are in fibreglass-reinforced composites. Renault considered that a structure made entirely from sheet aluminium would give a 40% weight saving over steel, but total costs would be too high. An aluminium front end would be 40–60% more expensive than the current Clio, assuming a production rate of 1800 vehicles per day. The aluminium front end, largely built from extrusions, was 30% lighter than the production Clio, whilst combining good stiffness, impact resistance and acoustic capabilities. Indeed, impact resistance was better than for the steel body: in a head-on crash at 50 kph, caving-in was 20% less than for the steel Clio, while the cabin 'survival space' remained entirely intact.

With the composite front end and floorpan, the MOSAIC project was more innovative. Again, the composite solution performed well compared with the reference steel Clio vehicle, with stiffness, endurance and NVH levels up to specification. Some questions remain over impact performance because composite structures absorb impact energy by crumpling. Indeed, composites can, if properly designed as components, absorb twice as much energy as steel on a weight-for-weight basis. In the 50 kph head-on test, the cabin remained intact but caving-in was 7% higher than for the steel Clio.

An important production advantage in composites is parts consolidation, well demonstrated by the MOSAIC project. The entire MOSAIC composite chassis (floorpan, front end, front body panel and rear wheel housings) consists of only 13 separate parts, compared with about 100 parts if built from steel (i.e. a 90% parts reduction), which generates important secondary effects in terms of tooling cost, time to market, and production logistics. The composite front end showed similar levels of parts consolidation, though critical adhesive lines had to be mechanically reinforced with rivets in the assembly process — incidentally giving an advantage as it could enable the structure to cure more slowly at room temperature without distortion as the vehicle progressed along the assembly line.

The Innovative Manufacturing Initiative

The Innovative Manufacturing Initiative (IMI) is a UK-based programme which addresses a number of sectors, one of which is land transport. Sectors were chosen on the basis of overall economic importance, and on the scale and pace of change required.

In terms of the automotive sector, the IMI is more than a technology programme, and does not seek to deliver discrete goals in the way that the PNGV does. Rather the intention is to identify key 'enabling' technologies and business processes that will underpin the success of firms into the next century. The IMI has three main themes of research and development:

- vehicle telematics and related systems
- alternative, lightweight vehicle structures
- supply chain management

The intention is that academics should join with industrial partners; industry will pay for the cost of the academic partners (payment can be in kind), while matched funding will be available from the IMI. An equally if not more important output of the entire process is the creation of dynamic networks of interrelated contacts as a key mechanism of technology diffusion and the integration of diverse technology streams. However, the IMI suffers from the 'risk averse' funding culture in UK government circles created by the 'value for money' climate, and the projects so far selected for support appear strongly orientated towards large, already well-advanced, and short- to medium-term projects.

Collaborative R&D in Japan

Japan does of course have a considerable record of major collaborative R&D programmes, and a history of close government–industry relations which would appear to make a firm foundation for an initiative in the realm of automotive technology. Yet it is apparent that there is no Japanese equivalent to the PNGV or the EU Car of Tomorrow. There may be a number of reasons for this:

(a) Japan has a powerful, advanced and cost-effective steel industry; in comparison it has no indigenous aluminium production and a very weak plastics industry. Thus there is little in the way of industrial promotion of alternative vehicle materials.
(b) The vehicle manufacturers in Japan have considerable expertise in very small vehicles which are relatively environmentally efficient.
(c) The relatively short product cycles in Japan should allow new technologies to be integrated more quickly than in other countries.
(d) The vehicle manufacturers in Japan have mainly given their attention to the continued refinement of steel bodies and to alternative engine technologies. It is indicative that the first production application of high-pressure direct injection petrol engines has come from Mitsubishi (see Chapter 5).
(e) Japan has a very well-developed electronics industry and, perhaps in consequence of this, has concentrated on the application of electronics systems to improve vehicle performance.

In this context, it is not surprising that the focus in Japan for government and industry collaboration has been in IHVS (Government of Japan, 1996).

3.8   CONCLUSIONS

In both the EU and the USA, traditional regulation of the car through mandatory standards is seen to have failed in some important respects, and the new collaborative R&D initiatives thus represent a new approach to reaching the desired goals.

Importantly, these R&D programmes are more than a mechanism to enshrine ecological principles into industrial change. They are seen as vital to ensuring the competitive survival of the automotive industry in the respective regions, and pioneering new ways of working, as well as meeting core environmental goals. One interpretation of these new ways of working is that government, through collaborative R&D programmes, is seeking to 'signpost' technological trajectories and thereby reduce for companies the risk and uncertainty that the technological transformation of the industry entails. The automotive sector is thus a key case study on whether international and collaborative action can create the mechanisms by which ecology can inform and underpin global restructuring.

However, collaboration is also a tool for achieving consensus. First, it defines a technical consensus on the nature of the problem and the most 'acceptable' means of resolution. This in turn provides the basis for a political consensus, vital if the proposed research is to be funded sufficiently. Collaboration is also bounded, especially in the case of the PNGV. That is, the definitions of who is allowed to participate in the collaborative effort, and in what capacity, are themselves deeply political. In this way, collaborative R&D programmes are a means of building strategic agreement as to future change. On the other hand, most of the debate surrounding these new R&D programmes has been conducted outside the public domain. It has been conducted by those party to collaboration, with little opportunity for wider debate.

From a more critical perspective the renewed interest in 'big science' collaboration between government and industry has distinct similarities to military R&D programmes. Indeed, the dire threats which accompany the PNGV and the Task Force appear as an echo of earlier arms races, with the sentiment that 'we must do this because they are doing this', and at least in the US it would appear that the previous beneficiaries of the Cold War who have seen the justification of their existence eroded by the changing global political climate are strongly involved in these new collaborative efforts (Wells, 1996b).

For the automotive industry, changing regulation and collaboration show that 'business as usual' is not a viable option. Core technologies in body construction and powertrain are under threat. It is to these issues that we now turn.

4 Budd and the All-Steel Body Paradigm

The revolution wrought by Budd brought technological spin-offs in its wake as surely as any other breakthrough in design, from the cannons of Crecy to the Sputniks in space.
(Leonard Setright: quoted in Grayson, 1978)

4.1 A BRIEF HISTORY OF CAR BODY DESIGN: THE ALL-STEEL BODY AS A SOCIO-TECHNICAL PARADIGM

The car can be traced back to Baden-Wuerttemberg in Germany in the 1880s. What made it possible was the development at that time of light, relatively fast-running internal combustion engines, pioneered by Otto and his disciples at the technical college in Karlsruhe. These were for the first time suitable for fitment to a simple chassis in that they were light and relatively powerful. Chassis design had also developed at this time. Essentially, car chassis design developed from two different traditions: the horsedrawn carriage or coachbuilding tradition and the bicycle tradition, the latter still being relatively young at that time. So we see a number of apparently unrelated technologies being brought together to enable the creation of the automobile.

Wood was the basic material of horsedrawn vehicles, often with a steel or steel-reinforced frame. These techniques were carried over to the motor car and can be seen in Gottlieb Daimler and Wilhelm Maybach's early four-wheeler Daimler designs. The bicycle, however, had helped the development of brazed steel tube and bent tube technology, leading to a much lighter and stronger structure than was possible in wood. This tradition is reflected in Carl Benz's early Patentmotorwagen. Thus in the world's first two motor cars with internal combustion engines we see these two competing technologies.

In terms of bodywork, these early cars were limited to seats and some basic protection from thrown up mud, usually in leather or wood, although bent metal appeared on some vehicles early on. Early pioneers of a unitary body-chassis design, such as Latil in France in 1899, were generally ignored and thus the car became established as a powered rolling chassis onto which a non-structural or semi-structural body was fitted. Bodies consisted generally of a timber (usually ash) frame, onto which steel or aluminium body panels were fitted: a 'composite' body. In some cases, lighter materials were used for body panels, such as the

Weimann type of a reinforced leathercloth material. Thus we have three competing cladding materials from an early age: steel, aluminium and 'composite' materials.

Early dissatisfaction with the durability of steel and the cost of aluminium led many to return to or stick with wood, especially for luxury coachwork (Reise, 1921, as quoted in Strassl, 1984: 57). The same ash frame was then covered with thin softwood or plywood panels. Walnut was found to be particularly suited to rounded panels. However, considerable time and labour input was required in order to steam and bend the wood for such applications. It occurred to some people within the industry that the two major subassemblies — chassis and body — could be integrated, with the following benefits:

- reduced weight
- reduced height, as people would not be seated on top of a chassis
- potentially simpler (hence cheaper) construction

Probably the first true monocoque since Latil in 1899 was a German cyclecar, the Grade Reibradwagen, of which around 1000 were built between 1921 and 1926 (Eckerman, 1989: 55). Other contenders included the French Voisin GP car of 1923, which led to the 'Laboratoire' production car, of which no more than 10 examples were built in 1924 (Courteauld, 1991). Voisin and his body designer, Noel Noel, used their World War I aircraft-building experience to design essentially an aircraft fuselage on wheels. A bath-tub-like construction of structural timber-reinforced aluminium panels was used on this car to produce a true monocoque (see below).

Aluminium was also used by the German aerodynamicist Wunibald Kamm, who built an all-aluminium monocoque conceived with ease of production in mind (Eckerman, 1989: 53, 55). The project was supported by the Schwaebische Huettenwerke (SHW) and three prototypes were built, while BMW considered putting it into production as its first car. In the end, BMW opted for the Austin Seven-based Dixi instead.

Another contender was the Lancia Lambda, launched in 1922 and on sale in 1923. This was more similar to the modern semi-monocoque or semi-unitary construction in that it had no chassis; instead it used a three-dimensional body structure carrying powertrain, suspension, etc. Onto this, exterior panels and separate wings/running boards were attached. Unlike the Voisin or the SHW, this was a true series production car and Lambdas were even used as taxis in Italy, although they were equally at home in rallys and on the race track as the design provided the car with excellent road manners and performance due to the stiffness and low weight resulting from the semi-unitary construction.

At the same time, the volume of cars built increased dramatically as a result of the introduction of mass production by firms such as Ford, GM, Citroën, Fiat and Morris. Traditional body construction was just too labour-intensive to adapt to mass production. In addition, styling entered car design from the 1920s onwards and the 1930s saw the advent of streamlining and aerodynamics, requiring much more complex shapes. These shapes could be achieved by hand in the aluminium-panelled luxury cars of the time, but reproducing them in mass production was not feasible.

Another problem was the painting process. Paint could take days or weeks to dry. Black was in fact the fastest drying paint, hence Henry Ford's legendary insistence on this colour as the least disruptive to his mass production process (in fact, any dark, high-pigment colour was available from Ford; see below). Baking improved the situation, but carried the danger of the ash frames catching fire — a not uncommon event at the time.

All these problems were solved by the vision and efforts of an American, Edward Gowen Budd, who developed the all-steel welded body, thus establishing the present car body paradigm. Initially his techniques were used to build the bodies for fitting onto steel chassis and the first Budd client for a volume application was Dodge Brothers Inc. At the time, one of the main advantages of this technology was the fact that it allowed proper baking of the paint, thus reducing the drying process from weeks to just one day. This removed one of the main bottlenecks and allowed the mass production system to grow much more rapidly. In the US, car bodies switched from being 85% wood in 1920 to over 70% steel in only six years (Lee-Harwood, 1996).

Budd and others soon realised the potential of this technology, which coincided with the development of tubular chassis. These were first developed for serious production by Hans Ledwinka for the Tatra 11 of 1923. Tubular chassis provided a level of stiffness hitherto unknown and from this, as Strassl (1984: 68) points out, it was a small step to start thinking in terms of a steel welded monocoque. The reason why it took so long for the monocoque concept to develop was summarised by Strassl (1984) as follows:

1. It was not known that the main stresses on the bodywork are torsional.
2. Weight reduction and fuel consuption were only an issue for very small cars.
3. Most cars were open tourers and did not benefit from a rigidity-enhancing roof.

In fact, Ware (1976: 70) states that in 1927 only 46% of all new cars registered in Britain were saloons. However, by 1931 this figure had doubled to 92%, largely due to the advance of the all-steel body through firms such as Morris (see below). The Lancia Lambda was a clear — and successful — attempt to adapt the requirements of greater torsional stiffness to an open tourer body. This was partly achieved at the expense of very small door openings. Only a fully enclosed body style allows one to think of a car body in terms of a large tube with holes cut out for windows and doors — the basic concept of a unitary body structure.

Nevertheless, the monocoque body took several decades to spread. Early examples were the Citroën 'traction avant' of 1934, the Opel Olympia of 1935 and the VW Beetle. The latter used a reinforced floorpan to act as a rudimentary chassis, a technique also used on the Citroën 2CV of 1948. It was not until the 1950s that the monocoque steel body became properly established. One of the reasons for this was that it was not until this time that the spread of mass motorisation and the resulting need for mass production made this necessary.

One of the last popular cars to use a chassis was the Triumph Herald, and derivatives such as the Spitfire and Vitesse. This is indicative more of Standard-Triumph's perilous financial situation at the time than any conviction against the monocoque. Nowadays only certain specialist low-volume cars retain a separate chassis, although the Toyota Crown had a separate chassis until recently, making it probably the last volume production car to be so endowed.

4.2   E. G. BUDD, FATHER OF THE MASS-PRODUCED CAR

Henry Ford is usually credited with the development of the modern mass production system in the car industry. This view is summarised by Freund and Martin (1993: 61) as follows: 'mass production of the auto was made possible by Henry Ford's combination of the assembly line technology with the division of labour scheme of Frederick Taylor'.

On closer observation, Ford's contribution is more limited. Henry Ford introduced the moving assembly line and was among the first to divide up processes into smaller units, more amenable to automation. He also contributed significantly to improving the logistics of the production process. However, as Williams et al. (1993) and Raff (1991) point out, his main contribution lay in the machine shop (see also McKinlay & Starkey, 1994). This enabled him to streamline component production. His innovations were such that few suppliers could keep up with the rate of expansion at Ford and this forced him into very high levels of vertical integration. This in turn allowed a much closer control over the component and materials flows inside the plant, and allowed economies of scale to be achieved in major subassemblies such as engines, gearboxes, axles and wheels.

However, the main bottleneck remained in the production of bodies. In terms of a modern assembly plant, Ford's legacy is primarily in final assembly. All the other essential processes in a modern assembly plant — namely press shop, body-in-white and paint shop — owe little to Ford. The introduction of these essential building blocks of modern car production and the cause of its high entry costs are due to another American, E. G. Budd, and his Austrian partner, Joseph Ledwinka — a relative of Tatra's Hans Ledwinka (Margolius & Henry, 1990: 122–3). The legacy of these men is usually overlooked.

It is important to realise that Henry Ford improved the building of an Edwardian vehicle. The Model T was not designed for mass production. It was designed before Ford introduced the moving assembly line. Ford's mass production was adapted to the Model T, which benefited from some improvements during its 20-year production life, but no radical redesign. In no way, then, was the T designed for mass production. Surprisingly enough, its successor, the Model A, was not all that much better, at a time when many of Ford's competitors were looking for more radical solutions.

Even modern commentators on Ford's innovations have often been confused by some of this. Karel Williams et al., in their book *Cars* (1994), point out — with some surprise — that Ford actually offered the Model T in a large number of variants, despite the widespread belief that mass production works best when all products are the same: 'In the myth of Ford, the aphorism about "any colour so long as it's black" serves as a metaphor for a company policy of restricting consumer choice. [in fact] The company's general policy was to win sales by extending consumer choice' (Williams et al., 1994: 99–100). As we saw earlier, black was the fastest drying colour at the time, hence the choice, but it is true that different chassis lengths and a wide range of different bodies were offered. 'And the secret of Ford's subsequent success was his ability to stretch a robust basic design so as to accommodate rapid changes in market requirements' (Williams et al., 1994: 100).

However, Williams et al., though noticing this anomaly in perceptions of mass car making versus the reality of Ford's extensive range of variants, do not seek to explain it and thus do not point out that this has little relevance for modern car production. Yet Williams et al. (1993), together with Raff (1991), provided a useful contribution to the Fordist debate by explaining that the exact nature of Ford's innovations were centred around his machine shop, i.e. not bodywork. The explanation of why Ford could offer considerable variation in specification is simple, however, and lies in the basic technological approach used by Ford. Ford built his Model T on a separate chassis. This technique allows considerable flexibility, especially in the possibility of mating a range of different body styles to a single basic chassis design. Any Edwardian vehicle maker had the same flexibility in terms of product specifications as Ford, and many used it in a similar way to build a range of variants on one basic design. This was illustrated much later by the Triumph Herald, Spitfire, Vitesse, GT6, etc., which all shared the same basic chassis.

The secret of Ford's success in offering a wide range of variants was therefore to a large extent due to the fact that he predated Budd in his technology. The problems and advantages now associated with mass production start with the adoption of the Budd technology paradigm. This was entirely based on the technology of pressing steel sheets, which itself was only developed around 1860 (Grayson, 1978: 352), about a decade before Budd was born. The technique was developed in New England and used in the first instance to make pieces for clocks, locks and other general hardware applications.

In relation to the universal acclaim of Ford's achievements, it is surprising so few people have heard of Budd. Edward Budd was educated as a metal-worker and served an apprenticeship as a machinist at the Smyrna Ironworks. He then moved on to the machine tool industry in Philadelphia, but joined up with his friend Thomas Corscaden when the latter received funding to develop his design for a pressed steel pulley, intended to replace the standard cast iron version. This American Pulley Company diversified into pressings for rail-car seats for Hale & Kilburn, who had a virtual monopoly in this area and who rapidly poached Budd.

Here he was set to work on developing pressed steel replacements for various cast parts. He played a key part at a time when the once-invincible cast iron industry was virtually being replaced by pressed steel. At the same time, the pressed steel industry itself was taking its technology further by being able to press ever thinner sheets with more accuracy. One of the first firms to recognise the potential in cars was Hupmobile, who approached Hale & Kilburn about this. Budd, who felt there should not be a wood piece 'as big as a toothpick' in a car body (Grayson, 1978: 353), was excited by this project.

They developed the Hupmobile model 32 for the 1912 model year. The bodies for this, the first all-steel production car, were made by Hale & Kilburn and shipped to Hupmobile in Detroit for painting and trimming. As mentioned above, the higher baking temperature now possible — due to the absence of potentially flammable timber — reduced the paint drying time from up to several weeks to only one day.

In 1912, Budd resigned from Hale & Kilburn and set up his own company to concentrate on the development of all-steel bodies. He was fortunate in that many of his former co-workers followed him, especially Joseph Ledwinka, the chief engineer to whom most of Budd's key patents were awarded. The main challenge initially concerned the welding technologies needed, and early attempts led to serious distortions in the bodies. Large body-framing jigs were developed, together with new welding techniques, while the pressings and die design also presented major problems. In fact this is still regarded as a 'black art' in press shops today.

Early orders came from Buick, Oakland and Willys-Overland; however, volume production did not start until the Dodge brothers — formerly senior Ford officials — decided to use steel for the body of their new car. In 1914, Dodge Brothers Inc. ordered 5000 all-steel tourer bodies, followed by 50 000 in 1915. Budd and Dodge developed a very close 'open book' buyer–supplier relationship, which benefited both (Grayson, 1978: 357). Dodge did not seem to attach a lot of publicity to the novel construction, so despite this revolution in body technology, the market was barely aware of how different Dodge cars were.

This is much like the Renault Espace in the 1980s and 1990s, where many owners are unaware of the different body construction, indicating, perhaps, that as far as market perceptions are concerned, a radical change in body construction need not be as risky as many manufacturers think. Opel, on the other hand, when launching the Olympia at the 1937 Berlin Motorshow, showed one of the cars with its body cut open in various key areas to show the all-steel body construction (Eckermann, 1989: 69). Perhaps the German car buyer was already more interested in the technology of his vehicle than his American counterpart.

A steel wheel activity, using a Michelin licence, was added in the early 1920s, by which time Willys, Dodge, Ford, Studebaker and luxury producer Wills Sainte Clair had become major customers. The way Ledwinka's patents were framed allowed Budd considerable control over steel body technology. In fact, from 1922 onwards, Budd had a monopoly in both welded steel body technology and

Table 4.1 Citroën Production, 1919–29

Year	Production volume (units)
1919	2 500
1920	20 200
1921	9 200
1922	22 000
1923	36 150
1924	67 200
1925	61 300
1926	36 300
1927	73 600
1928	63 400
1929	95 600

Source: Schweitzer (1982: 16).

concept (Schmarbeck, 1989: 45). As a result, licensing of the technology became an important part of the business, whereby Budd personnel usually acted on a consultancy basis in the setting-up phase.

Outside the US, one of the early converts was Andre Citroën, who visited Budd in 1923 and rapidly took on the technology at a royalty rate of US$5 per body (Grayson, 1978: 361). Citroën claimed that the adoption of Budd technology allowed him to increase his daily production rate from between 30 and 50 to between 400 and 500 (Schweitzer, 1982: 11–19). This is indeed borne out by Citroën's production figures, which show a sudden increase from 1923 to 1924, as illustrated in Table 4.1.

This technology also enabled Citroën to develop its revolutionary Traction Avant model, launched in 1934. Renault also adopted the technology, but only threats of legal action from Budd made him pay licence fees on the Juvaquatre model. In Germany, a joint venture, Ambi–Budd, was set up, which was involved in work for BMW, Adler, Hanomag, NSU, Auto Union, Opel and Ford. In 1925, William Morris visited Budd, which led to the formation by Morris and Budd of the Pressed Steel Company of Great Britain Ltd. Budd systems were first installed at Cowley and used for the Morris Cowley saloon of 1927. Other licences were taken out by Fiat, Simca, Peugeot, independent press worker Carel Fouche in France, Piaggio, Volvo and ZIS of the Soviet Union.

The spread of Budd's mass production during this period can be illustrated by the fact that while in 1933 US production represented 72.9% of all cars made worldwide, only five years later, in 1938, this had been reduced to 65.6%, and yet overall production worldwide had increased from 2 158 175 in 1933 to 3 041 948 in 1938 (Table 4.2). The difference was made up by increased production in the countries where the main producers had recently adopted Buddist practices. The fact that France shows a decline is due to market factors as well as the fact that the impact here came in the 1920s, as we saw earlier for Citroën.

Table 4.2   Car Production Worldwide, 1933 and 1938

Country	1933		1938	
	Units	%	Units	%
USA	1 573 512	72.9	2 000 985	65.6
UK	220 775	10.2	341 028	11.2
Germany	92 226	4.3	276 804	9.1
France	163 770	7.6	199 750	6.6
Canada	53 849	2.5	125 081	4.1
Italy	32 000	1.5	59 000	1.9
USSR	10 252	0.5	26 800	1.9
World	2 158 175	100.0	3 041 948	100.0

Source: Eckermann (1989: 74).

Table 4.3   The Car Parc in France, 1894–1983 (units)

Year	Parc
1894	200
1900	3 000
1914	107 000
1919	93 000
1930	1 000 000
1939	2 000 000
1945	910 000
1983	20 300 000

Source: Delerm (1986: 12).

This growth is also reflected in parc figures. Table 4.3 shows the total cars in use in France over time. Although it took some 35 years to reach the first million, the second million was achieved in only nine years. However, the true leap in numbers clearly appears after World War II, as is shown in Table 4.3.

In the process, Budd and Ledwinka redefined the car body in terms of a limited number of key subassemblies, namely cowl or scuttle/firewall, side panels, roof and rear panel. These formed the basic body-in-white onto which the hinged panels were hung. The chassis was still separate in this concept. The ability to press single side panels involved developing welded blanks, as the rolling mills were unable at this time to produce the width of sheet required. This technology was also developed by Ledwinka in the mid-1920s. Other innovations involved die production with the help of a mechanical engraving tool developed by Joseph Keller, whom Budd happened to meet on a train.

4.3 WHAT REVOLUTION?

One of the side effects of the Budd system is that it allows far less flexibility in terms of face-lifts or minor body changes. As Ware (1976: 69), comparing it to traditional practice, puts it:

> The ash frame provides the shape of the car, while the metal panels provide the strength ... Because of this simple construction it was much easier for a manufacturer to change the style of his bodywork when up-dating it, as there was not nearly as much expensive equipment involved in re-tooling, one of the major problems with present-day manufacture.

On the other hand, the Budd system has proved particularly amenable to automation. In the 1930s, it could still be claimed that car body construction was the most labour-intensive process in the car assembly plant:

> Grundsaetzlich muss festgestellt werden, dass wohl in keinem anderen Teilgebiet des Kraftwagenbaus so viel Handarbeit geleistet werden muss wie im Karosseriebau. [Basically it is the case that no other area of vehicle construction involves as much manual labour as body construction.]
> (Motor und Sport, 1935 in Strassl, 1984: 63)

However, by the late 1980s, this was by far the most automated part of the process, leaving final assembly as the only area where manual labour is still prominent. This susceptibility to automation has also meant that the Buddist areas have been particularly rewarding in terms of cost reduction. It is this scope for cost-cutting through Budd all-steel technology that is the true revolution of mass production in the car industry. It is this that has made the product affordable enough to make mass motorisation possible and it has had little to do with Henry Ford. The strength of Ohno at Toyota and the Toyota Production System or TPS (Womack et al., 1990; Toyota, 1992) has been that they have been able to recognise the tremendous scope for cost reduction, waste reduction and streamlining of logistics provided by the Budd paradigm.

These cost-cutting simplifications in body construction have not always been equally economical for the aftermarket. The once virtually universal bolted-on rear wings — still used on the VW Beetle, Citroën 2CV and Saab 96 — have virtually disappeared, making accident repair of such items more expensive. Over time, bolted fixtures have given way to welds, which are much cheaper in production though more expensive in aftermarket repair (Strassl, 1984: 61).

In order to understand the economics of modern mass car production it is therefore vital to understand the impact of Budd and Ledwinka. Redefined in this way, away from Fordist analysis and towards a 'Buddist' technological paradigm analysis, it is clear that the lean production revolution (Womack et al., 1990) is not, after all, so revolutionary. The major stepping stone from craft to mass production was provided by Budd, hence its adoption around the world in the wake of Budd. Citroën, Morris and other European followers all had Budd licences and by the 1960s the Buddist paradigm had become well nigh universal. This has not changed.

The lean producers of Japan, most notably Toyota, still work within the Budd technology paradigm and are, as a result, subject to the same basic economies of scale as every other volume car maker (cf. Rhys et al., 1993). They have managed — through the TPS — to optimise the system and thus cut costs, but ultimately they have not changed the paradigm. It is not surprising, therefore, that in the end they come up against the same problems as other all-steel car makers anywhere in the world.

What is significant is that the Budd company itself has not lost its spirit of innovation, and rather than sticking doggedly with the all-steel paradigm, like most of its customers, it has developed expertise in various alternative body materials, most notably plastic composites.

4.4   DEFINING THE BODY STRUCTURE

In view of the confusing terminology used in the present rush to look for alternatives to the welded steel monocoque, a brief review of the different chassis and body types may be helpful.

Ladder frame

The term 'ladder frame' refers to the most common type of frame used on early cars, although some cars use it to this day. The ladder frame consists of two I-beams, U-beams or box sections running the length of the car, normally from the front bumper to the rear bumper. Shorter cross-sections connect the two main beams. This system is also still used on nearly all trucks. A refinement developed just before World War I was the cross-bracing of this structure with a large X-shaped section. This cross-bracing dramatically enhanced the torsional rigidity of the frame.

Backbone frame

The backbone frame was popularised by the inventive Austrian engineer Hans Ledwinka on his Tatra 11 and 12 of the early 1920s, and later on the landmark 77 and 87 models of the 1930s. Modern Tatras, both cars and trucks, still use a variation on this system. This system consisted of a large-diameter round section tube running the length of the car in the middle. The wheels were attached at each end by means of swing axles, while the propshaft ran through the tube. Later Tatra cars, from the 77 of 1934 onwards, had the engine at the rear, but retained the swing axle suspension. This type of pure backbone chassis is now rare and apart from Tatra is only employed by the Austrian Steyr-Puch Pinzgauer off-road vehicle, as used by the British Army among others.

A modified version of the original backbone chassis is also found. This consists of a square section 'tube' or large box section built up from sheet metal or tubes in a spaceframe fashion. This type is used on the Lotus Esprit (box section) and on TVRs (tubular), among others.

Tubular spaceframe

Spaceframe is often used these days for modified monocoque types. However, the original spaceframe is a tubular spaceframe in which the powertrain and suspension are attached to a complex network of tubes welded together to create a kind of cage. The body is fitted on top of this. This construction was popular on Italian performance cars of the 1950s and 1960s. It led to the development of the 'Superleggera' construction by Touring of Milan, in which a very lightweight non-structural, often aluminium, bodyshell was lowered on top of the spaceframe with minimal attachments. One famous example of the type is the Maserati 'Birdcage', so named because of its tubular spaceframe; another was the Alfa Romeo Giulia Tubolare Zagato.

The tubular spaceframe combines strength and stiffness with low weight, but has a number of disadvantages. First of all, it is very labour-intensive to build; manual welding of the tubes is normally the only option, making it only suitable for expensive low-volume cars. The second problem is that structural damage is quite common after even relatively minor accidents, although more modern versions use sacrificial subframe sections at the front and rear to protect the main spaceframe. One modern example is the German Isdera Imperator supercar, which uses a square section tubular spaceframe. Several UK kitcars also use tubular spaceframes. A modern interpretation of this type is the Opel Maxx concept car, which uses aluminium extrusions to build a spaceframe, although this is halfway to becoming a perimeter frame.

Perimeter frame

The perimeter frame consists of a large beam forming a modified O-shape when looked at from above, around the outside of the car. This is a very rare type of construction, but is used on some recent lightweight electric cars and may become more common as a result. Thinner sections are usually welded onto the perimeter frame to make door/window frames and to hold powertrain and suspension. This way a perimeter frame can become a kind of external cage to which all other parts are attached, and in most cases this structure is actually visible. The 'holes' in the cage are filled in with panels or windows to keep out the weather as required. One recent example of this type of construction is the French Excel range of electric vehicles, although it departs from the pure perimeter frame in that it has an additional chassis for carrying the battery pack. Another example is the Opel Maxx, mentioned above as a mixed form.

Monocoque

The monocoque, unitary body or unibody design combines the functions of a chassis and body in a three-dimensional structure which carries powertrain, suspension and all hang-ons. It is the standard structure for most cars made around the world in mass production processes today. However, in many cases the system is better described as semi-monocoque in that most of these bodies carry additional non-structural outer skin panels. In a sense, therefore, the so-called monocoque is the chassis, while the skin panels are the remains of the traditional separate body. In a true monocoque the outer panels are structural, so the actual structure is visible. There have been few examples of a true monocoque, although the Jaguar E-type came close. It had a central 'tub' which had stressed outer panels. The engine was attached to a subframe which was attached to the front bulkhead. Only the large bonnet, the doors and the bootlid were non-structural.

Just as few modern cars are true monocoques, the Audi A8 does not use a true spaceframe. In fact, on closer inspection, the Audi uses a modified semi-monocoque. However, it does illustrate that in their search for alternatives, manufacturers are increasingly moving away from the existing body paradigm. Some of the basic types outlined above and mixtures or hybrid forms of these may well re-emerge.

4.5 THE DESIGN PROCESS

The design of cars is a process that has been subject to considerable change over the past 10–15 years, as this is increasingly regarded as a strategic tool in the competitive environment. Shorter lead times in design have become desirable for two main reasons:

● competition with the Japanese four-year model cycle in the popular segments
● the need to offer a new model regularly in order to keep the attention of the market

Nevertheless, in Europe a long product cycle is still considered appropriate for luxury cars in order to boost residual values and give a quality image. Much of the market feels that if a Mercedes or Volvo stays in production for a long time, it must have been right to start with (see Chapter 8). For the manufacturers of these products, these long cycles are essential if they are to reach the minimum economies of scale to justify their investments in the model and make a profit over the lifetime of that model. On the other hand, some models, such as basic A segment cars (e.g. Mini, Citroën 2CV, Renault 4) tend to have much longer model cycles, as do 4×4 sport utilities (Land Rover).

A new model is rarely all new. Many parts, especially powertrain and suspension components, are often carried over from the previous model. If a new segment is entered, these items can be shared with other models in the range. What is considered increasingly valuable in the market is less the technological changes under the skin, invisible to most customers, but more visible changes, which implies bodywork changes. As a result, body styling and flexibility in body engineering have become more of a competitive tool.

The traditional design process basically goes through a number of stages, which can be summarised as follows:

- Conception: what sort of car is going to be made for which segment, what engines, how much carry-over, how many components can be shared with other models, etc.
- Styling: this phase consists of a number of processes:
    — rendering; rough styling sketches establishing the basic shape
    — some of these are developed in more detail adding practical considerations, such as some material choices
    — model-making: building three-dimensional representations of between four and twelve chosen designs to scale, usually 1/5th
    — one or two of these may be built to a full-size mock-up in styrene/clay with details still changeable
    — one variant is chosen
- Design sign-off: the decision (by the board) to go ahead and produce the final choice
- Prototype build, usually in two main phases:
    — running the prototype to try out the mechanical configuration and new components in the existing bodyshell; this is done before design sign-off in many cases
    — running prototypes in the final bodyshell
- Productionising: production engineers try to make the car suitable for production
- Pilot production

Body material decisions could be taken at different points during this process, depending on which parts are affected. Materials for the monocoque have to be decided at the conception stage; for example, aluminium in the case of the Audi A8, and steel + SMC/RTM in the case of the Renault Espace. However, for hang-ons, such as doors, bonnet, bootlid or rear hatch, and even wings, materials can be changed at any stage, even while the car is in production. This is one of the reasons why essentially similar models can be made side by side with different materials, such as the Volvo 740 (steel bonnet) and 960 (aluminium bonnet), or Renault Clio (steel wings) and Clio 16V/ Williams (thermoplastic wings).

Various changes over the past few years have allowed the fusing together of a number of these stages. Computer-aided design (CAD) has allowed various stages from rendering to model-making and even the production of a full-size mock-up to be integrated. Some of these systems allow a link up with a simple milling device in a computer-aided manufacturing or CAM-like system to a milling machine which will enable the production of a mock-up from styrene foam or clay.

Some firms have gone even further. As model-making and the building of full-size mock-ups take time and money, and as the end product is relatively intolerant of change, a more flexible approach has been sought. Renault is probably the most advanced in this respect. Since 1990 it has been using virtual reality software to enhance the quality of the CAD process (Bonnaud, 1995). This has enabled Renault to produce realistic looking images of their designs in different, even real settings to allow senior staff to assess a design without making three-dimensional models. A full-size projection of the proposal can be made on a large screen and the car can be shown in real urban or landscape settings, allowing a much more realistic assessment of a car than from a three-dimensional mock-up in a bare room with artificial lighting.

As Renault is a pioneer in this technology, much groundwork is still being done and at present the system is still very expensive. The 'Les Citadines' video, featuring a range of futuristic urban vehicles driving through a real urban setting and mixing with contemporary vehicles, took three people some six months to put together for a final viewing time of only four minutes. The final cost was FF2 million. However, this is offset by the fact that far fewer mock-ups (FF1 million for one full-size mock-up) need to be made, that the model stage can often be bypassed altogether, and that in future much time can be saved (Bonnaud, 1995). The system has been used successfully on a number of concept cars, notably the Racoon and the family of urban vehicles featured in the 'Les Citadines' video. The concept cars were only built after the whole design stage had been completed using this sophisticated software, and in the future, production cars will at least partly be designed by this method.

Throughout the industry, production engineers are also brought in at a much earlier stage to allow potential production problems to be ironed out before they become time-consuming and expensive. Increasingly, in an approach pioneered by the Japanese, firms now form multidisciplinary product teams which develop a new model through all stages from design to production. In this way, each interest group is represented at each stage and able to avoid problems later on. This system has allowed the speeding up of product development and a reduction in cost. This approach has become essential in a lean production environment. Chrysler has claimed considerable benefits from this approach, which they first used on the Viper, but also on the more important LN range.

Another time-consuming part of the process is the production of the tooling, especially the press dies for steel body parts. Following Japanese practices these are now ordered in basic form at the beginning of the design process, with detailed information supplied to the die-maker as the design is signed off. Several alternative body technologies allow quicker tool production than steel and this may give them an advantage in the future.

4.6 ATTEMPTS TO EXTEND THE ALL-STEEL BODY PARADIGM: PLATFORMS

Another approach manufacturers are following is trying to optimise the number of variants derived from a limited number of basic platforms. Here a conflict may arise as to the extent to which a limited platform strategy can deliver product differentiation while retaining economies of scale. Some firms have implemented considerable platform rationalisation over the past 15 years, mainly to reduce the number of platforms inherited through the merger with other companies. These measures have led to considerable cost reductions as the basic floorpan or platform is the single most expensive subassembly to engineer and develop in a new car. The floorpan or platform can represent as much as 40% of the cost of engineering a bodyshell. But if two models at the same production volumes can be based on an identical platform, clearly this is reduced to 20% for each bodystyle.

Platform rationalisation has perhaps been most striking at Fiat and PSA, as outlined in Tables 4.4 and 4.5. This leaves the Fiat group with a dramatic reduction from 12 platforms in 1980 to 10 in 1985 (including the Croma-Thema, part of the Type 4 joint venture, which included the Saab 9000 and the Alfa 164, then not yet part of Fiat; we could therefore count this as half a Fiat platform, reducing the total to 9.5 platforms), and on to 7 in 1990. By 1995, the total was down to 6.5, including the Ulysse/Zeta joint venture with PSA as half a platform. In other words, the total number of platforms was reduced by half. The most ambitious platform range is the Tipo, which also led to the Tempra and Dedra, as well as being used as the basis for the Bravo/a — introduced in 1996. Alfa used this platform to derive the 155, as well as the 145, 146 and the GTV and Spider models, introduced in 1995.

At PSA too, the rationalisation results from the integration of a number of product ranges from Peugeot, Citroën and Chrysler-Simca. Again we should count the joint venture vehicles Peugeot 806 and Citroën Evasion as only 0.5 of a platform. In this case the number of platforms amounts to 10 in 1980 and 1985, despite some detail changes. However, by 1990 the number is halved to 5, while the move into the MPV segment only adds half a platform to reach 5.5 for 1995.

These rationalisation measures have led to significant cost reductions for both companies, making them much more competitive in the market. In addition to the product development savings, the companies need to sustain fewer platforms in production and this further reduces their costs.

Unlike Fiat and PSA, VAG actually increased the number of platforms it sustained. In the 1970s, Audi and VW models were built on the same platforms as follows:

VW	Audi
Polo	50
Golf	80/90
Passat	100/200

Table 4.4   Fiat Platform Rationalisation, 1980–95

1980	1985	1990	1995
126	126	Panda	Cinquecento
127	127	Uno/Y10	Panda
128 Ritmo	Panda	Tipo	Punto
		Tempra	Barchetta
	Ritmo	Dedra	
131	Regatta		Bravo/a
	Delta	Croma	Coupé
132	Prisma	Thema	Dedra
		164	Delta
X1/9	Uno/Y10		Tempra
		Delta	155/145/146
124 Spider	Croma		GTV/Spider
	Thema	33	
A112			Croma
	X1/9	Spider	Kappa
Beta		75	164
MonteCarlo	A112		
			Y10
Gamma	(Arna)		
			Ulysse
Alfasud >	33		Zeta
Giulia	Spider		
Giulietta	75		
Alfetta	90		
	6		

Note: we have included the Alfa Romeo range from 1980, although Alfa was not incorporated into the Fiat Group until the late 1980s. Alfa's Spider, which was phased out only recently, was built on the old Giulia platform, which also formed the basis for the Giulietta and Alfetta platforms, which in turn formed the basis for the 75, 90 and Alfa 6 platforms. The ill-fated Arna was a Nissan platform, fitted with an Alfasud powertrain. It should also be noted that the 126 is in fact still produced in Poland in small numbers for the local market, that the Uno is now also built in Poland, as well as Brazil. The Cinquecento is also built in Poland. The table excludes Innocenti, Maserati and Ferrari, which are also part of the Fiat Group.

During the 1980s, Audi was more and more allowed to differentiate itself from VW, and by the early 1990s these companies used separate platforms. One of the reasons for this was VW's decision to move to a transverse engine layout, while Audi retained its lengthways engine layout, making the Quattro system easier to engineer. However, the VW group has now decided to revert to a more integrated platform strategy in order to reduce its very high cost base. VAG has announced that future models for all ranges — VW, Audi, SEAT and Skoda — will all be built on only four different platforms. By 2000, the range will therefore look as shown in Table 4.6.

Table 4.5   PSA Platform Rationalisation, 1980–95

1980	1985	1990	1995
2CV	2CV	AX	Saxo
Mehari			106
Dyane	GSA	BX	
			205
GS	BX	XM/605	
			ZX/306
CX	104/LNA	205/309	
	Visa		Xantia
104/LN	Samba	405	406
Visa			
	205		605/XM
304			
	305		806/Evasion
305			
	505		
504			
	604		
604			
	Horizon		
Horizon	Solara		
1307/8/9			
2 Litres			

Source: CAIR.

Table 4.6   VAG Platforms for 2000 (introduction dates)

A	Polo, Cordoba, Ibiza (1999), VW Piccolo (1997)/SEAT Marbella (1997), Felicia
B	Golf 4 (1997), Vento (1998), Golf Coupé (1999), Concept 1 (1998), Toledo (1997), Audi A3 (1996), Audi TT/TTs (1998), Skoda A+ (1997)
C	Passat (1996), A6 (1997)
D	A8 (+ Isotta Fraschini)
+	A1/A2
JV	Sharan, SEAT Alhambra (1996) (+ Porsche Varera)

Source: CAIR.

In fact, as can be seen from Table 4.6, there are more than four platforms. First of all there is the shared Sharan platform, built in a joint venture (JV) with Ford at Auto Europa in Portugal. This counts as 0.5. In addition, there is the planned A1/A2 '3 litre car'. This is a small lightweight car which is planned for a fuel consumption of no more than 3 litres/100 km. It is currently envisaged as using an extruded aluminium spaceframe structure not unlike the Audi A8, although the panelling material is as yet unknown. This would then give VAG 5.5 platforms, if the spaceframe cars are counted as platforms. The A8 may derive some incremental volumes from the revived Isotta Fraschini cars built in Italy from 1996 in very small numbers.

4.7 ULSAB 1: THE PORSCHE PROJECT

The Ultra Light Steel Auto Body (ULSAB) project builds upon an earlier collaborative research effort, using Porsche, to develop steel material usage optimisation in a traditional vehicle body design. The ULSAB steel optimisation in Phase One was itself based on an earlier project by Porsche. This first project generated some useful results and is worth detailing here.

Steel optimisation on a traditional monocoque

The initial Porsche project took an existing (though dated) US car body as a starting point, and sought to identify incremental changes that would decrease weight without compromise to other performance attributes. Porsche quantified the weight-saving potential as shown in Table 4.7.

Some of the savings identified could be achieved in a current series production vehicle with relatively low cost changes. However, the majority would need to be implemented at the design stage because the savings would impact upon the design of tooling. However, the study was based on the notion of optimisation of current technology and materials. That is, the theoretical 30% potential weight saving could be achieved without significant new investment or alteration of existing manufacturing technology — though of course areas such as laser welding would probably demand new investment by the vehicle manufacturers.

Table 4.7 Identified Weight-Saving Potential (% of Weight Saved Over the Benchmark Body-in-White)

Item	Contribution to weight saving (%)
High-strength steel substitution	5.0
Flange size reduction	2.5
Carry-over parts reduction	2.5
Correct choice of material gauge/type	2.5
Welding/process changes	6.0
Design for greater formability	5.0
Component integration	6.0
Achievable potential (realistic)	15–20
Theoretical potential	30.0

Source: Hamm (1993).

Table 4.8   Weight Reduction Potential on the Previous Generation Ford Taurus

Item	Taurus production model (kg)	Weight reduction possible (kg)	%
BIW only	275.8	50.6	18.35
Closures only	62.0	13.0	20.97
Subframe	23.3	3.6	15.45
Front rails	17.2	1.8	10.47
Total	378.3	73.8	19.51

Note: an additional 4.8 kg is saved through redesigned flanges.
Source: AISI (1994).

Table 4.9   Main Sources of Weight Reduction on the Ford Taurus

Source	Contribution (%)
Mild steel gauge optimisation	6.7
Material substitution (HSS and BH)	4.3
Tailored blanks	12.2
Redesign	76.7

Note: HSS = high-strength steel; BH = bake-hardening steel.
Source: AISI (1994).

In a study on behalf of Ford (US) and the American Iron and Steel Institute, the Taurus model four-door mid-size sedan (by then quite a dated platform) was analysed by Porsche Engineering Services. Their findings in terms of possible weight reduction using existing technology are shown in Table 4.8.

An interesting aspect of this project was the extent to which weight saving was achieved by redesign of the structure as opposed to material substitution or tailored blanks (Table 4.9). In this project over 50 parts were eliminated from the design by consolidation, and an estimated US$40 cost saving on the body-in-white realised.

Porsche also undertook a design exercise to optimise the body structure from a 'clean sheet of paper'. This theoretical exercise showed a potential for savings of 126.1 kg on the main body-in-white (45% saving over the baseline vehicle) and 13 kg saving on the closures (21% saving), giving a total weight reduction potential of 139 kg or 41% over the previous generation Taurus. This exercise illustrates how far steel materials and vehicle design have improved in the period between the R&D for the previous generation Taurus and the present, but the Taurus design examined was over 10 years old.

An interesting by-product of investigations into steel welding techniques is that laser seam welding offers the twin advantages of reducing steel content while increasing body stiffness (ASI, 1994). Increasing the frequency of spot welds on the vehicle structure has relatively little impact on stiffness, but does improve fatigue loading (an important factor in design). However, more spot welds mean increased costs (fixed and variable), and also increased distortion which may give rise to more re-work. In one study by British Steel, weld-bonding was substituted for spot welding alone wherever possible, leading to a 25% reduction in the number of welds, while torsional stiffness of the body increased by 8.2% (Mathews et al., 1993; Mathews, 1994).

4.8   ULSAB 2: THE STEEL INDUSTRY VISION OF THE FUTURE

The Ultra Light Steel Auto Body consortium is an intensive, multi-phase study intended to demonstrate the ability of steel to deliver a significant weight reduction in vehicle bodies with reduced cost, and without compromising safety, NVH and performance. It is funded by the leading steel producers worldwide and conducted by Porsche Engineering Inc., a North American unit of the German company Porsche AG. The ULSAB consortium project builds upon earlier work also undertaken by Porsche, as outlined above. The ULSAB is more radical, and represents the core of the collective response from the steel industry to the demands for reduced vehicle weight (Hughes, 1995; Martin & Peterson, 1995; Schneider et al., 1995). The companies in Table 4.10 are members of the ULSAB consortium.

Phase One of the ULSAB, which sought to demonstrate technological potential, has been completed, with the results announced in September 1995. The ULSAB vehicle does not actually exist; the entire process has been conducted on computer. Phase Two, which seeks to develop actual vehicles, started in 1996. Phase Three will see the transition of the design approach into volume vehicle production — if the ULSAB approach is accepted by the vehicle manufacturers. The budget for ULSAB Phases One and Two is about US$22 million.

Phase One of ULSAB started in 1994. Several existing body structures were used as benchmarks against which the project would compare its design. The structures selected were as follows:

- Honda Acura Legend
- BMW 5 Series
- Chevrolet Lumina
- Ford Taurus
- Honda Accord
- Toyota Lexus LS400
- Mazda 929
- Mercedes-Benz E Class
- Toyota Cressida

Table 4.10   Members of the ULSAB Consortium

Company	Country
SIDERAR SAIC	Argentina
BHP Steel	Australia
Voest Alpine Stahl Linz GmbH	Austria
Cockerill Sambre SA	Belgium
Sidmar	Belgium
Dofasco Inc.	Canada
Stelco Inc.	Canada
Sollac	France
Krupp Hoesch Stahl	Germany
Preussag AG	Germany
Thyssen AG	Germany
ILVA Laminati Piani SpA	Italy
Kawasaki Steel Corporation	Japan
Kobe Steel Ltd	Japan
Nippon Steel Corporation	Japan
NKK Corporation	Japan
Sumitomo Metal Industries	Japan
Pohang Iron and Steel Co.	Korea
Hoogovens Group BV	Netherlands
CSI	Spain
SSAB Tunnplat AB	Sweden
British Steel	UK
AK Steel Corporation	USA
Bethlehem Steel Corporation	USA
Inland Steel Flat Products	USA
LTV Steel Co.	USA
National Steel Corporation	USA
Rouge Steel Co.	USA
US Steel Group	USA
USX Corporation	USA
WCI Steel Inc.	USA
Wierton Steel Corporation	USA

Source: ULSAB (1995).

These are, in North American terms, mid-size sedans — though in European terms this size of vehicle represents a small part of the total market. Originally ULSAB selected 32 benchmark vehicles, but detailed information was only available for the nine listed above. The composite average weight of the body structures of these vehicles was 271 kg. The design brief did not include doors, bonnet or the bootlid — just the body-in-white. The benchmark also showed a static torsional rigidity of 11 531 Nm/deg and a static bending rigidity of 11 900 N/mm. However, ULSAB also sought to compensate for current developments, and therefore set the benchmark performance higher with the following standards:

Table 4.11   Basic ULSAB Package Layout

Body type	four-door sedan
Wheelbase	2700 mm
Overall length	4800 mm
Overall width	1800 mm
Curb weight	1350 kg
Passengers	five
Headroom front/rear	960/940 mm
Leg room front/rear	1420/1400
Cargo volume	425 dm^3
Engine type	V6
Drive	front

Source: ULSAB (1995).

Mass	250 kg
Static torsional rigidity	13 000 Nm/deg
Static bending rigidity	12 200 N/mm

From the benchmark vehicles, ULSAB generated a basic package layout of the generic vehicle. This is show in Table 4.11.

The ULSAB target was then to reduce mass by 20% (to under 200 kg) without compromise to torsional or bending rigidity. The result was a body that weighed 205 kg without compromise of the safety or rigidity performance of the body, and which according to the consortium would cost US$154 less to manufacture than most current designs, though no volume figure was attached to these claims. That is, the project delivered a 35% to 15% weight reduction (depending upon which of the benchmark vehicles is used for comparison), a 14% cost reduction and an average 69% improvement in rigidity.

Porsche emphasised the importance of treating the body-in-white in an holistic manner, as an integrated system. Initially the body was designed on computer as a finite-element beam model, and then as a finite-element shell model. While some weight reduction was achieved with reduction of gauges in certain areas, the most significant reduction was achieved by the new architectural design of the body. For example, the design tried but excluded

- structural fuel tanks
- structural seats
- engine compartment brace (strut brace)
- cast nodes on the A pillar base joint

Alternatively, design changes included

- removal of the instrument panel beam
- a laser-welded roof rail
- an extended cowl joint
- a weld-bonded door frame

Table 4.12 Comparing the ULSAB Body with a Standard Unibody

	IBIS Unibody	ULSAB
Cost (US$)	1116	962
Mass (kg)	270	205
Piece count	195	169

Note: the body-in-white includes the structure plus roof and quarter panels.
Source: Hughes (1995).

The final model used hydroforming, laser welding, spot welding, tailor-welded blanks, weld-bonding, sandwich (laminate) steel, high-strength steel, bake-hardening steel, and roll-forming. Eighteen tailor-welded blanks were used in the structure. Weld-bonding of the A pillar inner, outer and hinge pillar, of the rocker inner and outer, of the B pillar inner and outer, and of the quarter panel inner and outer improved torsional rigidity by 1.3%, and bending rigidity by 8.0%. Continuous laser seam welds were applied to the roof-rail joins with the quarter panels and A and B pillars, and also to the area around the front shock absorbers.

For example, the hydroformed roof rail shows how traditional stamped steel parts could be replaced. This generates some joining problems in attaching other body-in-white components with a one-sided method (in this case, laser welding).

ULSAB economic costs

While at this stage hypothetical, ULSAB used IBIS Associates, a consultancy company, to compare the cost of their structure with a contemporary body-in-white. A summary of this comparison is shown in Table 4.12.

This gives US$154 per vehicle cost savings over a 'standard' body-in-white — though note that this is not a cost benchmark with the nine vehicles selected to form the reference point for the study. With the glass bonded in place, the torsional stiffness of the vehicle was 19 056 Nm/deg, significantly above the target, while bending stiffness was 12 529 N/mm.

The cost savings were expected to accrue from reduced die costs arising from eliminating parts, secondary savings in reduced assembly investment and variable costs, and reduced spot welding. Additional costs are imposed by use of high-strength and bake-hardening steel, and by the investment demands of laser-tailored blanks.

4.9 CONCLUSIONS

Weight savings in incremental improvements in steel bodies are not likely to be enough. The ULSAB proposals, although they offer some attractive and low-cost solutions, are problematic because they will entail the adoption of new production technologies (hydroforming, laser welding, etc.) which undermines a key advantage for steel — that existing production technologies may be used (Wells, 1996c). Moreover, there is an implicit acceptance with ULSAB that the outer panels (high value for steel makers, a core area for vehicle manufacturers) are the most vulnerable to materials substitution. Furthermore, to gain the full advantages of the ULSAB approach, the vehicle manufacturers would have to make a clean start, consisting of a total redesign with no carry-over from existing models.

Despite continuous improvement in steel-making technology and in design and manufacturing processes to transform steel into vehicle bodies, and despite the significant cost per unit advantage for steel in high volumes, all-steel car bodies are too heavy and the production technology too capital-intensive and inflexible. We will discuss the implications of this in Chapters 8 and 9.

5 Powertrain: Internal Combustion and the Limits of Electric Vehicles

We seem to be on the verge of a technological revolution
(Shoichiro Toyoda: quoted in Treece, 1996a)

5.1 INTRODUCTION

Toyoda, chairman of Toyota Motor Corp., was speaking at the 13th International Electric Vehicle Symposium in Osaka in 1996 and he was referring to the advent of electric vehicles. If he is right, the days of the internal combustion engine as the universal automotive power unit could be numbered. Toyota, Audi and other vehicle manufacturers around the world are responding by launching a hybrid electric vehicle.

In Chapter 4 we argued that steel body construction largely determines the minimum economies of scale in the car industry. However, this is closely followed by the mass production of engines, as Rhys (1984) shows (Table 5.1). As with vehicle bodies, the question of scale in engine production is closely related to the character of the manufacturing technology.

And yet, in engines, as in bodies, change is also expected to hit the industry. In fact, most of the environmental debate has focused on the emissions from car engines. The performance of car engines in all respects has improved significantly over the past two decades: fuel efficiency, reliability, longevity and environmental

Table 5.1 Optimum Scale in Key Automotive Production Activities

Activity	Units per annum
Casting of engine blocks	1 000 000
Casting of other parts	100 000–750 000
Powertrain machining and assembly	600 000
Axle machining and assembly	500 000
Pressing of panels	1–2 000 000
Final assembly	250 000
Paint shop	250 000

Source: Rhys (1984).

performance have all improved. These improvements have derived from changes in material (for example, the switch from cast iron blocks and cylinder heads to aluminium alloys), from greater precision in machining processes, and especially from the application of electronic control systems. Despite this, the industry is under considerable pressure to make further improvements. The proposed solutions to the emissions problems of internal combustion engines have centred around a number of key areas:

- alternative fuels: CNG, LPG, alcohols, rapeseed oil, hydrogen, etc.
- alternative combustion engines: external combustion (e.g. steam, Stirling)
- alternative powertrain options: electric traction (e.g. battery, fuel cell)

Much work has already been done on alternative fuels, and alternative fuelled vehicles are in use all over the world. Brazil ran its famous 'Proalcool' ethanol experiment for many years, although it has come under pressure in recent years. Liquefied petroleum gas (LPG) is in use around the world in countries as diverse as the Netherlands, Italy and South Korea, while France increasingly regards it as a cleaner alternative to diesel. Natural gas is also increasingly popular, especially for urban bus use; vehicles using this fuel are in use in Canada, Australia and experimentally also in, among others, the UK, Germany, Sweden and the Netherlands.

In Sweden, Volvo delivered 140 alternative fuelled buses in 1996 alone (Svelander, 1996). These are powered by natural gas, biogas or ethanol. Rapeseed oil, of which there has been a surplus in the EU, is considered a viable renewable cleaner alternative to diesel. In fact, the range of alternative fuels for internal combustion engines is almost endless. Although each has its problems, hydrogen is considered to be the most promising in the longer term. The attraction of the alternative fuel approach within the context of this book is the fact that a move to alternative fuels does not require a redesign of the existing engine production systems and the existing investments are therefore safe.

In fact, even many alternative engines can still be built using existing technology. Recent innovations such as the Orbital 2-stroke, Mazda's Miller-cycle engine, and also external combustion alternatives such as the Stirling engine, are similar enough to existing petrol and diesel engines not to require a radical change in engine production technology. Hydrogen, theoretically one of the most appealing long-term alternatives, can essentially be burned in adapted versions of existing engines, although it can also be used in fuel cells and it therefore does not guarantee the long-term future of the existing production system.

The problems only really arise with electric traction. This requires a completely different powertrain technology, which is only obliquely deriveable from existing automotive technologies. It is notable that the required expertise in the relevant areas such as batteries and electric motors tends to lie with major suppliers to the assemblers, such as Robert Bosch, Siemens and Valeo, and not with the existing car assemblers, although some suppliers, such as Delphi, Denso and Magneti Marrelli, are closely associated with particular assemblers.

A significant change to electric traction would, therefore, radically affect the existing technological and investment base in engine production and R&D in the motor industry. Ealey and Gentile (1995) focus on the implications of a widespread introduction of EVs for the motor industry, and their conclusions are for change of a similar magnitude as we foresee from a change of body technology. We will return to this later. The pressure for change towards a greater number of electric vehicles comes at the same time as the pressure for change in body technology, and the implications for the industry are clearly going to be dramatic as two of its core technologies are under attack.

A wholesale move towards electric traction would render existing engine production systems obsolete. In addition, out of all these alternatives, it would have the strongest impact on the future of car body and chassis design; in particular, a switch to electric traction exposes the problems of overall vehicle weight which are currently obscured by the power delivered by internal combustion engines. For these reasons, we will concentrate in this chapter on the electric vehicle alternative. Any move to an increase in the number of electric vehicles is likely to have a greater impact on the basic design of the motor car than most of the other alternatives that are likely to be widely available over the next few decades. A significant move towards electric vehicles, as advocated in the US, for example, has serious implications for our existing transport infrastructure as well as for the design of cars themselves.

5.2 BATTERY ELECTRIC VEHICLES (BEVs)

Whichever way you look at it, the internal combustion vehicle (ICV) is remarkably efficient, especially considering its still considerable scope for improvement. Thinking of electric vehicles as replacing our cars for every application is unrealistic in the short to medium term. It has been put this way: 'Who would buy an expensive small car with a one gallon tank which took eight hours to refill?' (Waters, 1992: 115).

In this respect little has changed since the beginning of the century, when EVs competed directly with ICVs. In the US, for example, where 'electrics' were particularly popular, of all the 4200 cars sold in 1900, 38% were electric, 22% petrol-engined and 40% steam-powered (Shacket, 1981). However, in practice, electrics were used differently. The largest producers of BEVs in the US were Baker and Detroit Electric, whose marketing was aimed primarily at middle-class urban and suburban women drivers. The ease of operation (no crank start, no non-synchromesh gears to change) appealed to this group, and the way these customers used the vehicles suited the BEV precisely. The shopping trip, social calls, etc., kept them well within the maximum 80 km range of the vehicle at speeds of around 30 kph, while using them only during the day allowed for the overnight recharge. The battery pack lasted, on average, three years (Stein, 1966: 100). At the time, the industry was waiting for a breakthrough in battery technology. This has not changed 90 years on.

The second wave of interest in electric vehicles came in the late 1960s and early 1970s. Having started originally as an attempt to develop small low-pollution city cars, it received a boost from the first oil crisis when people began to look for alternatives in order to reduce oil dependency. The present, third wave appears different in that there is a stronger commitment throughout the industry, although the influence of impending Californian ZEV legislation should not be underestimated. Mom and van der Vinne (1995: 11) argue that the difference this time round is environmental legislation.

In terms of the overall environmental performance of BEVs, PSA (Peugeot-Citroën) has calculated that less carbon dioxide and other greenhouse gases are produced powering an electric vehicle, than are emitted from a petrol or diesel engined vehicle. Although sulphur dioxide emissions would increase (because the electricity is derived from fossil-fuelled power stations), the installation of flue gas desulphurisation equipment would reduce this considerably (Coventry Electric Vehicle Project, 1996: 5). Noise pollution is also dramatically reduced by BEVs. In an urban transport environment, EVs have a distinct energy consumption advantage. EVs do not use energy while stationary and this can mean an energy saving of some 70% in an urban context over ICVs. PSA argue that even taking into account a life-cycle energy picture, the net energy saving can still be 17%. In addition, most recharging would be carried out at night when electricity demand is low, allowing a useful load levelling effect for the generating industry (Coventry Electric Vehicle Project, 1996: 6).

5.3   WHO PAYS?

Initial thinking in California was that the requirements for zero emissions vehicles (ZEVs) market share would largely be met by households choosing an EV as a second car. In the US, more than 40% of the parc consists of households' second cars. Of these, 90% are replaced every five years. In the US in 1980, despite the spatially extensive nature of urban development, about 87% of car journeys were of 15 miles (25 km) or less, while 95% were less than 30 miles (50 km) (Shacket, 1981). In the UK in the early 1990s, only around 18% of households had two or more cars (see Table 5.2), although this has increased slightly since. In most cases

Table 5.2   Households Owning Two or More Cars in Selected European Countries (%)

Country	Proportion (%)
United Kingdom	18
Netherlands	11
France	25
Italy	20

Source: van Dijk & Cramer (1992).

these are second-hand and in most cases they are cheaper and smaller than car number one. To reconcile this type of market with the higher price of the BEV clearly presents problems: hence the attempts to find ways of reducing the cost of BEV ownership without recourse to out and out subsidy.

Even in California it has been suggested that the expensive and relatively short-lived battery packs could be owned publicly or by a utility which then rents or leases them out. BMW has suggested a similar scheme for its E2 BEV, whereby the customer would buy the car but lease the battery pack. However, more recently General Motors has been promoting a full leasing programme for the EV1 electric car, which cannot be bought outright. The programme was launched in December 1996 and involved no down-payment and US$640 a month for a 36-month lease in San Diego, Phoenix and Tucson. However, residents of the Los Angeles area only paid US$480 a month, due to local subsidies. The recharger is leased for an additional US$50 a month (Rechtin, 1996). Leasing or renting the vehicles is attractive, because consumers then do not have to incur the risks of outright purchase of an electric vehicle.

The higher purchase price is one of the chief problems usually associated with an EV. With higher volume production the cost of both vehicle and battery pack is likely to come down; however, a differential with the existing mass-produced ICVs is likely to remain for some time. The price problem is highlighted by the estimates made of the premiums needed over ICVs in order to make BEV production profitable, as shown in Table 5.3.

The problem is the comparative unit costs of producing EV and IC vehicles. Some estimates have been made of the volumes needed to close the cost gap but in truth these seem to be unrealistically low. Table 5.4 illustrates a range of estimates.

It is possible that in some of the richer and 'greener' European markets, a limited number of private individuals will be prepared to invest in electric vehicles. Indeed, a number of people in Denmark and Switzerland have already done so. No special infrastructural arrangements have been necessary in these countries. The owner plugs in for recharging at home overnight. Some houses have solar panels for additional power generation.

Table 5.3 Price Premium Estimates for BEVs

Source	Price premium estimate (%)
MITI estimate	200
MITI target	20
ENEA (Italy) suggested maximum	10–30
US proposed legislation	11
Panda Elettra	160
Chrysler TEVan	600

Source: Diem (1992a).

Table 5.4   Minimum Estimated Efficient Scale for EV Production

Source	Units/year
CARB	3 000
Marie-Jo Denys MEP	60 000
AVARE (European EV association)	40 000–60 000
Fiat	20 000
LEM Holding SA	25 000–50 000
PSA	50 000

Source: compiled from various industry sources.

More recent thinking in California and elsewhere has therefore focused more on the utilities as initial users. The electric utilities clearly have a vested interest in increased BEV use and have generally proved supportive; indeed they have taken a more pro-active 'technology forcing' stance. The value of large demonstrator fleets of EVs and their ability to build an EV market share relatively quickly are also important factors. Besides, a central fleet depot is easier to fit out with recharging, and dedicated repair and maintenance facilities than thousands of private homes. Finally, keeping a new technology out of the hands of consumers limits the dangers of destructive litigation if something goes wrong. This is a major factor in the US, which may stifle much future development.

5.4   WHICH PROBLEM DOES THE BEV SOLVE?

We should perhaps focus on the actual problem that has prompted the Californians and others to look again at EVs. The problem is simple: urban air pollution (in the 1970s it was oil shortages). The US Federal Clean Air Act and its amendments list a number of urban centres that require urgent action. Their local authorities therefore have an obligation to attempt to improve their local situation, without necessarily taking the wider implications into account. French sources suggest that the car is responsible for some three-quarters of urban air pollution (Robert, 1992). This is important as up to 80% of Europeans now live in towns and cities. Nevertheless, we are dealing with a local problem, caused by the volume of traffic and the attendant problems of congestion.

This has a number of implications. First of all, a local problem needs local solutions (see, for example, Have, 1993, for the case of Amsterdam). This is a feature of many environmental problems. Apart from such issues as global warming, ozone holes or contamination of the North Sea, few environmental problems are actually national or international. Most of them occur in a particular town or city. This has prompted towns, cities and regions all over the world to take unilateral action on such matters (Figure 5.1 and Table 5.5). This has happened from Southern California to Athens, from Florence to La Rochelle. In general, environmental pressure groups are also more powerful at the local

Figure 5.1 Towns and Cities Worldwide have Taken Action to Restrict Access for Cars

level, enabling a harnessing of democratic forces behind a particular issue such as a city centre car ban (cf. Topp & Pharoah, 1994).

Italian cities have been among the most prominent in this respect. An example is Turin, which now claims to be the European leader in electric car promotion (*Made in Fiat*, 1996). In 1996, a park-and-drive scheme was introduced which

Table 5.5 Local Restrictions on Car Use in Europe

City	Restrictions
Bologna, Turin, Rome, Milan, Athens	Restrictions on car use on certain days/hours
Bergen, Oslo	Road pricing for entering city centre
Berne	Very limited number of unrestricted through roads; 30 kph speed limit on all other roads
Zurich	The 'Zurich Model' restricting the car, promotion of public transport and cycling
Hanover, Luebeck, Aachen	Complete car ban in city centre; promotion of cycling and public transport
Munich	Planned road tunnels dropped in favour of new underground lines; increase in cycling facilities

Source: CAIR.

allows drivers to park their IC car and hire a BEV for use in the city centre. Twenty Fiat Panda Elettras have been made available for hire at LIT3500 per hour for up to three hours and LIT5000 per hour thereafter (*Made in Fiat*, 1996). Payment is by means of an electronic prepayment card. The range is 50 km and the vehicles are exempt not only from the city centre car restrictions, but also from many parking regulations. The scheme is run by the city authorities, Fiat Auto, AEM, the local electric utility, and ATM, the public tram operator. The cost of setting up the scheme was around US$1.3 million.

This may require a rethink in the UK, which has seen increasing centralisation of decision-making during the past two decades or so. However, in reality, many local authorities have already taken such unilateral action on environmental matters, albeit on a small scale and mostly before the wave of centralisation. The bus lanes in Norwich were the first in the country, and have since been copied in many other urban areas. Pedestrianised zones all over the UK, the London Lorry Ban, etc., are other examples, while a number of cities, such as Manchester, have opted for light railways. Although a light railway is a relatively expensive option, it has the advantage of showing a clear commitment on the part of the local authority to tackle the transport problem. However, as Whitelegg (1993: 159) points out, 'The main flaw in Britain's transport policies is the enormous weight of a centralised and dogmatic London-based administrative machine.' The damage done by 17 years of centralisation in the UK will take a while to undo.

Another implication of the fact that EVs are a response to an urban air pollution problem is that there is not really a strong case to be made for using EVs outside such polluted urban areas, where the problem either does not exist or is not serious enough to warrant such a radical change in motorised personal transport. Especially as the change is to a mode of transport which may have low emissions at point of use, but which is less energy efficient over all than an ICV and may therefore ultimately cause more pollution elsewhere.

This has implications for the BEV infrastructure, because it means that this too can be confined to urban areas only. This suddenly makes the whole thing much more feasible and much cheaper. It also makes the BEV a much more realistic alternative to the ICV, as it only needs to replace the latter in its function as urban runabout, to which it is not necessarily most suited in the first place. However, this would burden the individual household with high expenditure on a vehicle whose use is severely restricted. There are always going to be a number of individuals who either because they can afford it or because they like the idea are prepared to make such an investment. But the vast majority of people — and it is always the majority that causes the problem — will not be interested. The point is that 'The benefits accrue to the general public, but the burdens are on the individual customer' (John Wallace of Ford Motor Co., quoted in *Automotive News*, 12 October 1992).

5.5 PRIVATE OR PERSONAL TRANSPORT?

The social gain is not taken account of by the individual. The externality therefore has to be internalised. The only way to induce consumers to opt for electric vehicles and help meet the targets set for ZEV penetration in California, for example, would be to introduce substantial incentives to EV buyers. If private individuals are to be induced to act in the interest of the community, the community should pay at least some of the cost. One incentive suggested by Madsen Pirie (1990) of the Adam Smith Institute, a right-wing think tank, is for UK company car incentives to be limited to electric vehicles.

There is another solution and that is for the BEV parc to be owned and operated by the local authority. Private operators could take over, but the problem is that they have no control over infrastructure decisions, so they would be at the mercy of the local authority; not attractive for any private business. A partnership is obviously possible, which gives us two options:

- local authority owned and operated BEVs
- BEVs owned and operated by a private company in partnership with the local authority

Whichever scenario is chosen, the local authority would have a major input in the local BEV infrastructure, even if the actual vehicles or even parts of the infrastructure hardware are privately owned. However, it should be pointed out that BEVs would also have to compete with public transport in the urban environment. As they would not solve the problem of congestion, widespread adoption of BEVs could further hamper public transport efficiency and postpone essential measures to restrict car use. Local authority input, on the other hand, would enable BEVs to be incorporated into an integrated transport policy at local level. This is a further attraction of this approach.

The average household would have access to electric vehicles for urban use, but would not be forced to buy one as a second car. In fact, one could envisage a scenario whereby households have a choice in the future between different types of vehicle. The present day multi-purpose car would be replaced by three more specialised types from which a customer could chose: the privately owned BEV, the privately owned hybrid or the privately owned lightweight ICV (cf. Nieuwenhuis et al., 1992: 43). In addition, public transport provisions could be enhanced and would benefit from being integrated with the personal public transport system provided by publicly owned BEVs. For their personal transport needs, the consumer could have a series of choices between private, public or personal public modes.

5.6 URBAN GOODS TRANSPORT

For commercial vehicle applications, the situation appears more straightforward, as large-scale operators will be able to take control of infrastructural requirements themselves. Battery weight and limited range are less of a problem for urban delivery vehicles and several fleet users in countries like France and the US run test fleets. Leaders in this field are, naturally, the power generating companies such as EDF in France. PSA Peugeot-Citroën has been particularly keen to sell to local authority users and as well as around a dozen local government customers in France, they have sold vehicles to China Light Power in Hong Kong. As an incentive, the French government allows companies buying BEVs to depreciate them within one year.

However, EV fleets are also run by Kodak and a number of postal services around the world, as well as, of course, many of the UK milk suppliers. In 1980 there were more than 60 000 milkfloats on UK roads, half the world EV parc; however, by the mid-1980s this number had halved and by the early 1990s numbers were down to 28 000 (Mom & van der Vinne, 1995: 18) and falling. The demise of the UK milkfloat has been caused by a combination of factors: changes to lifestyles, including a reduction in average per capita milk consumption; more dispersed urban populations; and the sale of cheap milk by supermarkets as a 'loss leader'. The vehicles themselves, however, have proven remarkably robust, reliable, and easy to maintain. In the future, more fleet users could be encouraged to opt for EVs. Urban taxis are an example listed by Gormezano et al. (1992: VIII-28), who also suggest vehicles in holiday complexes and local authority vehicles as early candidates for EV conversion.

5.7 INFRASTRUCTURE NEEDS

Existing BEV operations all involve fleets located centrally, recharged and maintained centrally, and always returning to base at the end of the day. The UK milkfloat system is a prime example. Electric golf carts and luggage transporters at airports and railway stations around Europe also come into this category, as

Table 5.6 BEV Infrastructure Requirements

1. BEV recharging stations/vehicle collection and drop-off point
2. Battery-pack exchange, storage and handling facilities (e.g. recycling)
3. BEV repair and maintenance facilities
4. Safety measures, training
5. Power-generating requirements

Source: Nieuwenhuis (1992).

do electric forklift trucks, AGVs and other EV elements of internal logistics systems. For EVs replacing cars, the situation is more complex.

In terms of the main road vehicle infrastructure, the electric vehicle is no different from existing internal combustion road vehicles; it can use the same roads, car parks, etc. The differences lie in the supply of energy. With the existing ICV fleet we use a decentralised nationwide refuelling network through the privately owned and operated petrol stations. This network is not necessarily suitable for BEVs, although Shacket (1981: 212) suggests that 'charging stations will begin as an adjunct to regular service stations'. However, recharging is still a time-consuming business which cannot be compared with refuelling an ICV. In fact, battery-pack exchange would be more comparable to fuelling an ICV, both in terms of concept (energy exchange) and timing (a few minutes), as outlined in Table 5.6.

BEVs can be recharged from the national electricity grid. Most models currently available can be plugged into the existing domestic supply for slow charging, for example. However, in view of their limited range it is felt that additional recharging provisions should be made available in public places even for privately owned and operated BEVs. In California, all new office car parks now have to include BEV recharging facilities (Gormezano et al., 1992: VIII-29). This means that vehicle energy supply will shift away from petrol stations to private houses, offices and car parks. Although some fuel stations could be retained as recharging points, an all-electric fleet would make petrol stations redundant. We mentioned above that rather than a long recharge, a swap of battery packs may be more efficient. Shacket (1981: 212) predicts that, 'Battery packs will be exchanged and owned by manufacturers. When the battery system has reached the end of its useful life, the manufacturer will recycle the components, minimizing the power required to build new batteries.' This may be a realistic way to go; however, it will require a high degree of standardisation if a range of different vehicles are likely to call in for a battery pack swap, as would be the case with widespread private ownership of BEVs.

Again, having a centralised body such as a local authority as a BEV operator in each area would greatly simplify this type of operation, as BEVs are likely to be limited to one model. Nevertheless, special measures are needed for storage of the bulky and potentially hazardous battery packs. Special handling facilities for exchanging these very heavy items are also required (e.g. forklift trucks).

Ultimately, all battery packs need to be exchanged as they come to the end of their useful life. GM estimated that the replacement cost of a battery pack for its Impact electric sports car would be around US$1000 every 30 000 km. This figure assumes recharging to cover the 30 000 km between necessary exchanges. In fact, GM division Delco Remy offers an exchange battery pack for the GM EV1 for between US$1500 and US$2000, although this is unlikely to represent the true cost.

Watson et al. (1986) suggest that the recharging infrastructure could be provided at a manageable extra cost. They assume private ownership and operation of BEVs and recharging at home overnight. Their study also assumes that most users would park their car off-road for recharging at home, although they do consider the implications for those without such facilities. They assume a total conversion of the UK car fleet to battery-electric operation. Energy supply would be provided by home charging and charging in public places ('opportunity charging', 'biberonnage'), as well as battery exchange. They estimate that the annual infrastructure costs would amount to £478 million, £290 million and £207 million for advanced lead acid, nickel–zinc and sodium–sulphur battery systems respectively (at 1980 figures). This is around 5–10% of the cost of the EV fleet. It does not include taxes.

Watson et al. (1986) recognise that their assumption of an all-electric car fleet is unrealistic before about 2010 at the earliest; however, some other elements could also have been considered in their study. They do not consider the effects of the total dismantling of most of the existing fuel station infrastructure for example, nor do they consider the possibilities of a publicly owned BEV fleet for personal use. In fact, the cost of a publicly owned personal public transport infrastructure is likely to be lower.

The power generating side is more straighforward. A European study group (Fabre et al., 1987) suggested that based on actual patterns of use around Europe, some six million cars (7% of the total European car parc) and one million local delivery vehicles (12% of the total European commercial vehicle parc) could profitably be replaced with electric versions without the need for an increase in generating capacity. The main reason for this is that most recharging would take place at night, when demand is low and therefore spare generating capacity is available. The result of such a change would be a cut in emissions by 2.5% and a reduction in energy consumption by the transport sector of 3.5% (cf. Waters, 1992: 119; Nieuwenhuis et al., 1992: 91).

Apart from the basic energy supply network, BEVs also require maintenance and repair regimes that are different from the existing ICV fleet. Special facilities and tooling will be needed, but most important is the special training and retraining of personnel involved in maintenance and repair both of the EV parc and of the special EV recharging infrastructure. Input from vehicle producers and electricity supply companies is essential in this area. However, it has been found (see Waters, 1992) that the maintenance costs of electric vehicles are actually some 30% lower than those for comparable IC vehicles. Ealey and Gentile (1995: 110) suggest that EVs would require half the maintenance of ICVs.

The new structures required for successful BEV operation require significant investment and organisational input. Much of this would be concentrated in clearly defined urban areas. During the start-up phase of this new infrastructure local authority involvement is therefore not only welcome but essential. As we will see below, this approach is already being taken in a number of places. Where a fleet of publicly owned or operated BEVs is provided for use by the public, special car parks may have to be provided for people from outside the city centre at the interface with the BEV. Alternatively or additionally, BEV pick-up points could be located to interface with public transport.

The question of safety is still often overlooked. Battery electric vehicles are inherently different from the existing parc, and dedicated safety legislation must be considered. Given the very high weight of the battery pack, for example, it would need to be very well secured in order not to pose a hazard for occupants and others. Batteries can contain dangerous substances which may be released or may explode in a crash, while the higher operating voltages needed for greater efficiency also bring with them the danger of electrocution if a short-circuit occurs in a crash (cf. Mom & van der Vinne, 1995: 54). Adressing this is vital, as one fatal accident with a small EV in these early stages could do untold damage to the fledgling EV sector (Mom & van der Vinne, 1995: 54). Similarly, we have seen in the UK how an ill-conceived design, such as the Sinclair C5, can set back the cause of EVs for many years (Figure 5.2).

Figure 5.2 The Sinclair C5 Set Back the Cause of EVs in the UK for Many Years

5.8   STANDARDISATION

On the technical side, standardisation is going to be an important issue. With EV schemes springing up all over the world — and the most active countries being the US, Japan, France and Italy — different standards are likely to evolve. Although this is not really a problem for the users and operators who limit themselves to a local area, it could prove a nightmare for the manufacturers expected to supply operators from Japan to California and Brazil to France. As mentioned earlier, especially if private ownership of BEVs is promoted, standardisation becomes essential.

One step in this direction may be a recharging connector developed by GM-Hughes and CALSTART in California. This allows charging by magnetic induction rather than metal-to-metal plug contacts and is said to be safer and more convenient. However, at present a monopoly exists for the supply of this system. The potential dangers involved in regular recharging by unskilled private individuals of privately owned and operated BEVs in private or public locations deserves further attention.

Various different approaches to battery packs are still found depending on the battery type used. As noted in Chapter 3, one of the most advanced areas of work within the PNGV concerns that which is seeking to develop new battery technologies, the US Advanced Battery Consortium (USABC). Thin-film lithium batteries in particular have been given considerable attention. Standardisation here would facilitate battery pack exchange schemes and allow exchange for a different battery type as developments proceed on this front.

5.9   BEV INFRASTRUCTURE IN PRACTICE

Some experiments designed to involve local authorities in the BEV infrastructure are now taking place in various locations around Europe. The French seem to be the most committed to this approach, the French government having earmarked some FF500 million for electric vehicle research. On 28 July 1992, PSA, Renault and EDF (Electricité de France) signed a framework agreement with the French ministries for industry and environment to create an electric vehicle infrastructure consisting of recharging and service facilities. The agreement specifically commits the companies to provide recharging stations along urban streets and in parking areas, as well as a battery-pack leasing and recycling system, and specialised service and repair centres.

By 1995, at least 10 urban areas in France were to be fitted out with such a basic electric vehicle infrastructure, although this plan has suffered some delays. A joint venture of PSA, Renault and EDF is carrying out the work in conjunction with the local authorities selected. These include La Rochelle, Chatellerault and Tours (all in western France), which have already introduced some pioneering schemes, although in all some 40 towns and cities have put forward proposals.

EDF is principally responsible for the infrastructure requirements. It already runs its own test fleet of more than 350 electric vehicles to gain real life experience. EDF has calculated that, based on an overnight recharging regime, it can supply power for up to 2 million BEVs in France without any increase in generating capacity (Polo, 1992). The recharging infrastructure will consist of three different elements (PSA, 1992: 9):

- Home outlets. Any conventional domestic 16 A socket can be used for a standard 8-hour recharge.
- Standard recharging stations. These will be located around the city in selected private and public parking areas. They use a conventional 230 V 16 A socket, but include interactive software to give handling instructions, and arrange payment by card. A safety system is also included which prevents charging in case of user error, vandalism or bad weather conditions. These units are intended for a normal recharge while cars are parked for prolonged periods, e.g. overnight.
- Rapid recharging points. These are designed to provide a fast emergency recharge when a vehicle has run out of energy. The 25 kW 10-minute rapid charge will allow another 20 km to be driven. These will often be located on garage forecourts or other convenient locations. They will also incorporate interactive software to communicate with the vehicle as well as the user, and to arrange card payment.

La Rochelle

The first practical trial was run in La Rochelle from January 1994. During 1993, around 40 recharging points were installed in La Rochelle, both in private locations and public places. The city agreed to reserve a number of parking places for EVs throughout the metropolitan area, to be fitted with recharging points by EDF. EDF also installed a number of rapid recharging points in selected service stations in the city. All these were used to recharge 50 converted Peugeot 106 and Citroën AX cars using Ni–Cad batteries. These were put at the disposal of selected individuals, carefully chosen to be as representative as possible of normal drivers. A cross-section of the private and business communities was represented. The vehicles, both cars and small vans, were leased at a commercial rate to all the users in order to ensure objectivity (Coventry Electric Vehicle Project, 1996: 12).

Some of the findings were particularly interesting to PSA. It was found, for example, that most recharging was carried out at home and at the place of work. The seven public recharging points were in fact little used (Coventry Electric Vehicle Project, 1996: 13). Initial worries on the part of the participants about range proved unfounded. In actual use, the patterns coincided with expectations in that 85% of all journeys were of less than 10 km, average weekly distance travelled was around 225 km, and the cars were recharged five times a week on average. Peugeot has made the 106 Electric available to the public in France and if successful in the longer term, Citroën may introduce the purpose-built Citela EV.

Tours

The La Rochelle approach is based on a system whereby a basic EV infrastructure is provided by EV manufacturers, public utilities and the local authority for the benefit of privately and business run EVs. The approach taken in Tours is slightly different. Here, PSA has linked with Via GTI (Generale de Transports et d'Industrie), a public transport company, and CEGELEC. From 1995, a system of 'personal' or 'individual public transport' was to be started. This system enables private individuals the rental of an electric 106 or AX by means of a standard banker's card. The vehicles are available from ten recharging stations in the city centre and have to be returned to one of these anywhere in the city after use. Although the system was to be operational by the middle of 1995 (PSA, 1992: 10–11), it suffered some delays.

Amsterdam: the Witkar experiment

There is a precedent for the Tours scheme, which operated in Amsterdam for more than 12 years. In the late 1960s, the Amsterdam hippy movement produced a political grouping, the 'Kabouters', which campaigned on several environmental issues. After two of their number were elected onto the local council, they managed to introduce the first stage of their plan for a new deal on urban transport. This 'white bicycle' plan involved a large fleet of bikes owned by the city of Amsterdam and painted white. These could be used by anyone at any time within the city. If you saw one, you could take it (a common practice anyway) and leave it when you reached your destination. This project had flaws in that white bicycles were soon found all over the country and even abroad. Copenhagen has recently introduced a similar scheme, although it involves a number of safeguards against this happening, such as a unique design of bike and a coin deposit system (McGurn, 1996).

Stage two was the 'Witkar' project, first conceived by Luud Schimmelpenninck in 1968. It involved small two-seater battery-electric vehicles of a curiously tall design. The vehicles had a range of 30–40 km and were to be kept at around 15 recharging stations, of which at least five were actually built. After a short experimental phase with two recharging points and five vehicles, the first recharging station opened in March 1974 (Coöperatieve Vereniging Witkar, 1974, 1975). The stations where vehicles were recharged via a roof-mounted contact were normally unmanned, with repair and maintenance taken care of at one central depot.

Use was restricted to members of the co-operative society that ran the project. A once-off joining fee of HFL25 (about £10) ensured membership. A special computer keycard was issued to members on deposit of another HFL25, which operated the Witkar and automatically charged your Witkar account — a world first. The fee for use was HFL0.15 per minute of operation. After use, the Witkar was returned to the nearest convenient recharging station. Membership reached a peak of 3000 in the late 1970s.

From the start, Witkar had to compete with the private car. The comprehensive inner city car ban that would have made Witkar viable was never implemented and for many years the scheme was kept alive by a small band of enthusiasts. About 25 Witkars were built in all and a new, three-seater version was planned in the late 1970s, but never saw the light of day. By 1986 the number of recharging stations had dwindled to three. The early computer system proved unreliable and regularly over- or undercharged customers. Modern systems are on the whole more reliable.

In 1986, a possible saviour appeared on the scene in the shape of Anglo-Swedish shipping magnate Arne Larsson, who planned to take on the project, commission 100 improved vehicles and build another 12 stations. Larsson had plans to introduce similar schemes in Frankfurt, Düsseldorf or Munich ('Lei', 1986). In the end the deal did not materialise and the Witkar project was halted in 1988. The whole project was remarkably cost-effective in that during its entire lifespan of more than 12 years it cost only around one million guilders (£380 000) to implement and operate.

This project was clearly ahead of its time and deserves to be revived. It shows that a city-centre-based system of individual public transport can work and can work at moderate cost. Combined with an integrated transport policy involving a car ban, improved public transport and comprehensive cycling provisions, it would no doubt last well beyond Witkar's lifespan. This highlights the point that emerges from this study: that the electric infrastructure is not only dependent on technological and economic factors, but at least as much depends on political and social factors. Rather than merely replacing existing cars with EVs (e.g. Watson et al., 1986), we should regard EVs as an opportunity to reform and update our transport infrastructure generally, both public and private as well as the relations between the two.

Coventry

Initially the Coventry scheme was to be run along the Tours model, but, in fact, the way it is now implemented is more like the La Rochelle scheme, with the exception that all the users are businesses. The scheme is run jointly by Peugeot Motor Company plc, Coventry City Council, East Midlands Electricity, PowerGen and the Royal Mail. Altogether the partners run a fleet of 14 electric vehicles, of which six are cars and eight vans; all are Peugeot 106 models. The project is monitored by the Energy Savings Trust, who will produce a full report on the experiment (Coventry Electric Vehicle Project, 1996: 4). The project's cost is around £400 000, part of which is funded by the Energy Savings Trust under the UK government's National Air Quality Strategy. The starting date was

November 1996 and all partners paid a commercial leasing rate for the vehicles. In addition, East Midlands Electricity was responsible for developing the recharging infrastructure in line with the partners' vehicles' duties.

Initiatives promoting urban electric vehicles have taken off in various other locations such as various smaller US communities and the town of Mendrisio in Switzerland. Japan also has several electric vehicle initiatives. In 1991, the Tokyo City government put 28 electric vans at the disposal of milk and rice distributors for a one-year experiment (Mom & van der Vinne, 1995: 20). In fact, Daihatsu may well be the world's largest EV producer at present, with total sales of 6000 since 1967, of which 1053 are road vehicles, although PSA may overtake it soon. In 1993, Daihatsu doubled its EV output to 800 a year, mainly an electric version of the HiJet minivan (Mom & van der Vinne, 1995: 18–19).

It is clear that the electric vehicle now enjoys a greater measure of support than it has ever done before. Its popularity is growing and this momentum in itself may ensure success. From a policy perspective, the critical issue now is to sustain and expand these initiatives so that alternative transport systems become embedded in the built environment.

5.10   DESIGN IMPLICATIONS

A move from IC to EV technology has a number of design implications. Many electric cars available at present are merely the manufacturers' conventional IC designs converted to electric power. In this way one can buy an electric Peugeot 106, for example, or an electric Fiat Panda. Although these are relatively light cars by contemporary standards, they are not purpose-designed for EV use. The energy density of petrol and diesel has allowed the vehicle manufacturers to engineer all cars without weight minimisation as a priority.

In the battery-electric powertrain, energy density is much lower, and the overall performance of the vehicle is much more sensitive to weight. Weight reduction therefore becomes a major consideration. GM has calculated that every 22 kg saved gives an extra mile of range (Mom & van der Vinne, 1995: 14). This has prompted many EV designers to produce a very lightweight structure for their vehicles as a means of compensating for the weight of the battery pack. Examples of this are the General Motors EV1, the BMW E1 and E2 prototypes, and the Horlacher-designed Saxi and Hotzenblitz small-volume EVs (Figures 5.3 and 5.4). A particularly interesting example of this is the PSA Tulip, which has a body structure consisting of only five large plastic panels (see Chapter 6).

It is this requirement for weight reduction, exposed most overtly in electric vehicles, that has produced a series of alternative body/chassis structure designs. A logical next step is the realisation that having engineered this lightweight structure merely to negate its elegance by adding a heavy battery pack is unappealing. Fitting the same structure with a small IC powertrain makes for a much more energy efficient vehicle all-round. In fact, BMW did this by fitting its E2 with one of its motorcycle engines. Another possibility is a hybrid powertrain

Figure 5.3 The Horlacher-Designed Saxi BEV

whereby the electric drive is retained, but instead of a large and heavy battery pack, a small battery pack and a small IC engine are added. Hybrids are discussed further below.

The second major impact of a transition to BEVs beyond the issue of vehicle (and component) weight is that of power consumption by accessory systems in

Figure 5.4   A Danish El-Jet BEV in Use in Heidelberg

the vehicle. Again, with the conventional ICV these sub-systems are relatively easy to accommodate, but in the BEV case energy conservation becomes of vital importance to the overall performance of the vehicle. Typical problem areas in this context include lighting and signalling; heating and air conditioning; power steering; electric motor-powered windows, seat adjusters, etc.; and radio and communications systems, including navigation.

5.11   HYBRIDS

In the early days of electric vehicles some attempts were made to prove the electric's ability to undertake long journeys, but these involved (illegal) recharging from 550 V overhead tram or trolley-bus lines. Even at that time, the hybrid was regarded by many as offering the best of both worlds. A hybrid-powered car has two different power systems: usually one is an internal combustion engine and the other an electric traction motor. Around the turn of the century, interesting hybrid production vehicles were made in some numbers by companies such as Lohner-Porsche in Austria (1898–1905) and Kriéger in France, who even produced an experimental gas-turbine electric hybrid in 1908. The Lohner-Porsche vehicle used hub motors and regenerative braking. However, these vehicles were considered too heavy and complex to compete with the petrol-engined car.

Nowadays, IC-electric hybrids are not uncommon in transport applications. Diesel-electric trains are in use worldwide, while most of the heaviest dumpers as used in mining and major infrastructure projects are also hybrids. There are two major types of hybrid: the parallel hybrid and the series hybrid. VAG has been pursuing the parallel hybrid route. Their thinking is of a dual mode vehicle that can run as a ZEV on electric traction in urban areas, while reverting to IC mode on unrestricted out-of-town roads. The advantages of this arrangement over a pure BEV are the much smaller battery pack and the increased range of the IC engine. On the other hand, running two powertrains adds weight and complication.

None the less, Audi is launching such a car, the Audi Duo, in September 1997 (*De Telegraaf*, 1996). The Duo, based on the Audi A4, uses a 1.9 litre direct injection diesel engine and a 21 kW electric motor, while both engines drive the front wheels via the standard five-speed gearbox. However, the car uses an automatic clutch which enables the electronic control unit to engage either diesel or electric drive, as appropriate (*De Telegraaf*, 1996). The range on pure electric drive is around 50 km; however, the additional weight is considerable, amounting to some 320 kg for the battery pack, although the additional weight of the electric motor has been reduced from 60 kg on the Audi Duo concept car of 1989 to a mere 22 kg. This initiative brings the hybrid drive car a lot closer to reality.

An example of a series hybrid is the medium-size city bus shown by Iveco in recent years. This uses a small Fiat car engine as the IC element. This is used purely as a generator, with enough power for the electric traction motors to move the bus. General Motors has developed its HX3 experimental series hybrid people-carrier, which uses a three-cylinder petrol engine to charge the 32 lead-acid batteries (Vauxhall, undated: 6).

The series hybrid also allows a revival of the gas turbine to be considered. After early experiments in the 1950s and 1960s by firms such as Fiat, Rover and Chrysler, this technology was abandoned in the past because of its high-speed requirement (60 000–80 000 rpm) and consequent turbine lag, making it sluggish when changes in speed were required such as under acceleration. However, the turbine is very efficient if run at a constantly high speed. In recent years electric generators have been developed that work at these high speeds and this has made the simple, light and compact gas turbine once again a candidate for automotive applications. Volvo's Environmental Concept Car (ECC) was presented as a gas turbine electric series hybrid in 1994, while Renault and PSA are involved in a government-backed programme to develop a new automotive gas turbine.

The motor industry finds hybrids particularly attractive as they allow it to go along the electric vehicle route without abandoning its expertise and investment in IC technology. Such an approach retains the 'personal mobility' appeal of the car, and is conducive to continued individual ownership of cars. However, the gas turbine would to some extent undermine this advantage as its technology is quite different from the existing piston engines.

5.12   FUEL CELLS

The fuel cell is based on the principle that certain chemical reactions generate electricity. A fuel-cell-powered vehicle would carry these substances on board and by combining them en route the electricity generated can be used to drive an electric motor. Excess energy can be stored in a battery pack, while regenerative braking can be used for energy conservation, as in a BEV. The advantages of this approach are the much reduced battery pack, the relative simplicity of the whole process, and the low emissions. Most firms are pursuing the hydrogen route. Here, hydrogen is combined with oxygen in the fuel cell to produce water.

Fuel cells were first seriously developed for the NASA space programme and although theoretically a very elegant solution, many recent examples have been very bulky. Much development is still needed, although a 1996 Toyota prototype based on an electric version of the RAV4 sport utility and named EVS-13 looks promising (Treece, 1996b). The EVS-13 uses methane from which the hydrogen is extracted, this is then bonded with a powdered alloy before being fed to the fuel cell. Toyota argued in favour of methane because of the existing methane (natural gas) delivery infrastructure. The company does not expect production versions before some time during the first decade of the 21st century.

5.13   IMPLICATIONS FOR THE CAR INDUSTRY

If we had to make a prediction, then hybrids are the most likely short- to medium-term solution, either with a petrol or, more likely, a diesel IC component. In the longer term, gas turbine hybrids are possible, overlapping with early fuel-cell vehicles. Battery-electric vehicles will form a growing niche for dedicated urban applications in the short to medium term. The main user groups are likely to be fleets operating in urban areas such as mail delivery, police and utilities.

In the very long term, electric vehicles have a major advantage in that they will allow us to reduce our dependence on fossil fuels, notably oil. There are a number of different ways of generating electricity, many of them (wind, solar) considerably 'greener' than fossil fuels. This makes a move towards electric traction in some form particularly attractive from an environmental and sustainability viewpoint. However, it is clear that any move towards a massive increase in electric vehicles could have severe repercussions for the existing car 'system', notably

- infrastructure
- engine production and development
- body structures

Ealey and Gentile (1995: 102) explain how 'the EV, if it becomes a mainstream mode of transportation, could change the entire auto industry, from design and development to sales, service and infrastructure'.

Like bodies, engines have come to be considered by the industry as one of their core technologies, adopted with the Budd system and the advent of mass production. Mass production allowed engines to be made economically in-house, while, as Ealey and Gentile (1995: 108) point out, 'The engine's secondary characteristics have also become salient sales points ... interlinked with the identity of the OEM that builds it.' They explain that there is little point in manufacturers developing electric motor production in-house as there is sufficient capacity and expertise in the relevant supply industry. They also argue that the resources released by abandoning IC technology could be deployed elsewhere (Ealey & Gentile, 1995: 110). These resources may be welcome in the new world of high-tech body technology and lightweight engineering with high electronics content. However, they also review the dramatic adverse effects of such a move for human resources requirements, dealer networks, the aftermarket (service requirements will be halved) and the existing suppliers.

The core business of the OEMs could be significantly reduced by the combination of a move towards alternative powertrain, especially EVs, and a move to different body materials. This could benefit the suppliers, but only those in relevant product areas and who can take on a large share of the product development in these areas with which the assemblers may be unfamiliar. Ealey and Gentile (1995: 113) summarise this as follows:

> Suppliers who survive in the new environment will be those that make the fundamental changes to their businesses that keep them relevant, in an era in which automobiles will have significantly different components than they have today.

This is remarkably in tune with the car makers' current strategy of devolving more and more activities up the supply chain to their components and systems suppliers.

Apart from the various advantages of electric vehicles, the US has an additional agenda in pushing for a wholesale move to EVs. The US Big 3 and their associates perceive themselves as having a considerable technological advantage in EV technology over Far Eastern competitors, particularly the Japanese. It has been suggested therefore that the US auto industry has chosen EVs as the last 'battlefield' against the Japanese (Mom & van der Vinne, 1995: 17) and that this is not the first time that non-technical considerations have become decisive in the introduction of a new set of technologies: in the 1980s there was the catalytic converter versus lean-burn debate in Europe (cf. Nieuwenhuis & Wells, 1994: 99). However, this time such non-technical considerations could lead to a paradigm shift with far-reaching economic, social and technical consequences. As Mom and van der Vinne point out, such a change implies a potential abandoning of some of the basic tenets of modern automobility, i.e. the culture that values top speed and acceleration as the inalienable rights of any motorist (Mom & van der Vinne, 1995: 18).

While the US may have the technical expertise for electric vehicles arising out of the skills and technologies developed under military and aerospace industries, it is not clear that the market and use conditions are as favourable as, say, in Europe. Not only does the US have a much more dispersed spatial structure and lower population density than many European and Asian countries; the culture of (ICV) car consumption is more deeply embedded.

Moreover, GM itself has referred to its EV1 facility as an 'experimental factory' in which the intention is to learn more about electric vehicle production processes. So the processes of developing the market will have to run in parallel with the process of developing the production systems.

6   Concept Cars: Visions of the Future

Our way of life is inflicting damage on the natural environment. A designer's key concern should be to modify the destruction of the environment.
                    (Robin Day, designer of the polypropylene chair: quoted in Withers, 1996)

6.1   INTRODUCTION

In Chapter 4 we explained that the existing body construction paradigm in the car industry is being questioned. In Chapter 5 we exposed a similar threat for the established internal combustion engine paradigm, primarily from electric vehicles. However, it would be unrealistic to assume that the car industry is unaware of these threats. Indeed, as we have shown, the automotive industry itself is the source of much of this alternative thinking. The industry recognises the threats and takes them very seriously; however, the inherent inertia in the existing paradigm makes the industry seem slow to act. One of the clearest indications of the industry's awareness of the threat is the steady stream of concept cars incorporating alternative powertrains and body technologies which the industry has been presenting in the recent past. These cars present visions of possible future automotive developments as seen by the industry.

6.2   VISIONARIES PAST AND PRESENT

Predicting the future has been as popular and as fanciful in the motor industry as in other walks of life. Every decade has seen the launch of a car that captured the imagination of forward-thinking individuals. In this way, the Voisin cars of the 1920s, Tatra 77 in the 1930s and Citroën DS in the 1950s came to symbolise technological, and with it often social, progress. However, other car makers went further and presented the public with 'dream cars' which were supposed to show the vehicles and lifestyles of a more distant future. The 1950s saw a mushrooming of such developments, with the US Big 3 presenting a series of future transport concepts.

Even the usually critical observers and historians of the industry such as Ralph Stein (1962: 312) were swept along: 'Isn't it inevitable that sooner or later cars will use atomic energy?' At the 1958 Paris Motor Show, the Arbel Symetric was shown, which was intended to run on spent nuclear fuel, which was allegedly used to power electric hub motors. Nothing has been heard of the car since

(Chapman, 1996: 10). In a sense the various French electric vehicle projects' cars do so indirectly through their use of electricity generated by nuclear power stations, although this is probably not what Stein had in mind. However, some of his other assessments were, with hindsight, more realistic:

> The passenger car of tomorrow will almost certainly borrow heavily from the sports car of today ... The channel-framed chassis is almost done for right now ... Within a few years the cast-iron engine will be as rare as the horse-collar ... The disc brake will become universal in a very few years, even on the cheapest Detroit car.
>
> (Stein, 1962: 306)

Other views Stein expresses in the book are perhaps more visionary than he may have realised: 'For the more expensive machine, the tubular space frame covered with an aluminium skin is almost certainly in the offing' (Stein, 1962: 306). We have indeed seen this on the Audi A8. However, some predictions have not quite materialised yet:

> ... there will be an increase in the number of rear-engined designs, as engines grow lighter in relation to the power they produce. Owing to the use of aluminium alloys it will become more practical to use more powerful engines in the rear without making cars unmanageably tail-heavy. Transmissions ... will also move towards the back, and what Detroit calls a 'trans-axle' (gearbox and differential in one unit) will be commonplace.
>
> (Stein, 1962: 306)

Although the last of these predictions has come about, it is within the context of a transverse FWD layout. Although mid-engined configurations have become reasonably popular for sports cars, its application has remained limited. Surprisingly, perhaps, Stein failed to predict the front-wheel-drive revolution.

Forecasting is clearly not easy. However, many of the predictions of the 1950s were fanciful as they also tried to predict future technological developments. The difference now is that the industry shows concept cars which embody existing technology, even if they are not necessarily production ready. In other words, these cars can be made and are often built as driveable prototypes. These cars show much more a strategic direction into which the producers think automotive development may or should go. This is strengthened even further by initiatives such as the PNGV in the US, which set the industry on a clear technological trajectory (see Chapter 3). As a result, predicting the future of the motor car is now easier than ever.

Fortunately, there are still visionaries willing to stimulate our imagination with pictures of the future. Rudolf and Robbert Das (1995) do this through drawings and they cover transport as a whole. Their cars of the future are more radically different, albeit more realistic than some of the 1950s proposals. They are more aerodynamic and use different seating arrangements. Powertrains consist of gas-turbine electric hybrid or fuel cells and electric drive. Suspension is hydro-pneumatic according to the Citroën system. Bodies are made mainly of composite materials.

However, they concentrate on how these cars interface with the infrastructure. Intelligent highways and alternative car-carrying systems using magnetic levitation feature in their vision of future transport systems (Das & Das, 1995: 69). Dedicated highways for heavy trucks also feature. The far-reaching automated driving of the intelligent highways increases the demand for driver-controlled leisure driving routes: the car as leisure object – a return to its roots (Das & Das, 1995: 70). Leisure vehicles include off-road vehicles (Das & Das, 1995: 132) as well as sophisticated articulated motorhomes (Das & Das, 1995: 136–7). For the developing world and urban environments, they propose greater use of human-powered vehicles (Das & Das, 1995: 144–5, 155).

The Das brothers take a 50-year horizon and many of their proposals seem perfectly plausible. Lightweighting is implied in their proposals, though not made into an overriding design concern. In a sense, enough technologies exist either in developed or embryonic form to paint realistic scenarios for the future of the motor car; what is perhaps more difficult is predicting the future of the industry. However, we will analyse first of all how the industry itself views its future products.

6.3 CONCEPT CARS: POINTERS TO THE FUTURE OF AUTOMOTIVE TECHNOLOGY

In order to analyse more closely the thinking within the manufacturing industry as to the future direction of car design, a review of such recent concept vehicles is useful. Concept cars presented by manufacturers as well as more specialised vehicle design and engineering companies can give a clear indication of a company's thinking. The one-off nature of such designs often means that they use plastic technology to construct vehicle bodies even where steel would be used in a production version. However, some of these concepts use alternative materials in a more deliberate way in both body design and construction. In these cases an analysis is more valid.

Over the past two decades, several concept cars have attempted a radical reappraisal in terms of car design with a view to making the car radically more efficient. In practice, little of the thinking embodied in these vehicles has thus far reached the market. However, these are more than mere show cars; they represent a clear indication of the thinking within the company that creates it and can often give a pointer to its future strategy.

We give below a brief description of some of the more relevant concept cars that illustrate new ways of thinking in terms of body materials and construction, as well as powertrain. It is significant that many recent concept cars represent a radical departure from current mainstream car technologies. Concept vehicles are the pointers to these technology choices; they narrow the pool of technologies from which manufacturers are likely to choose and as such are a valuable indication of future trends. Sources used are the press releases and brochures from the respective manufacturers, as well as Janicki (1992) and others where indicated.

6.4   EUROPEAN CONCEPT CARS

Mathis VL333 (1946)

The Mathis VL333 was not intended as a concept car. In fact, it was intended for serious production. Several fully working prototypes were constructed and displayed at the 1946 Salon de Paris in the hope of support from the government, which had to allocate scarce resources to any new car project. This car probably was too radical for the politicians and civil servants, as the specification below will illustrate. The specification was determined by a detailed design brief, embodied in the original name, Voiture Economique Legere (VEL); it was capable of a fuel consumption of 3 litres per 100 km, and fitted with three wheels and three seats (hence 333). The celebrated French aerodynamicist Andreau acted as a consultant for the body design, which helped achieve astounding aerodynamics for a practical car. The specification included (*La Vie Automobile*, 1946: 132; Rees, 1995: 53–4):

- Aluminium chassis and body; three wheels, two at the front with wide track; independent front suspension; front-wheel drive.
- Total weight: 440 kg.
- Cd: 0.22.
- Engine: 707 cc; horizontally opposed twin cylinder, water-cooled, two radiators.

It is interesting for our purposes in that it shows that even at this time aluminium was considered a feasible option for weight reduction in a small economy car. Within a year this thinking led to the first aluminium car in series production, the Panhard Dyna.

Porsche LLC (1973)

The Porsche Long Life Car project was commissioned by the German Federal Ministry for Research and Technology in the wake of the Club of Rome's report, *The Limits to Growth*. The team aimed at a useful service life of 20 years and 300 000 km, and much of the work focused on materials choice. The aims proved to be reasonable within a costs–benefits analysis. The LLC research developed two alternative materials scenarios, one retaining a steel body and one using aluminium. The savings and costs in each case were found to be as shown in Table 6.1; all figures are based on 1970s practices.

The steel body used double-sided hot dip galvanised steel for durability. It was this research that prompted Porsche to introduce fully galvanised bodies for its own production cars; a move later followed by Audi, which assembled the 924 and 944 models for Porsche. In addition to the body-in-white, the Porsche study also recommended upgrading of the other systems, particularly the electrical system — the source of many reliability problems in older cars. Engine and transmission life were increased by, among other measures, the use of

Table 6.1 Porsche LLC Research Findings

Item	Steel (%)	Aluminium (%)
Life-cycle energy saving	5	20[a]
Materials saving	55	65
Added cost at 1000/day	22	30
Additional labour	7	17

[a]Using recycled aluminium.
Sources: Bott et al. (1976), Nieuwenhuis & Wells (1994: 160–1).

hydrodynamic clutches, breakerless ignition systems, and improved air and oil filtration.

Porsche concluded that a gradual introduction of LLC technology would result in only a small and gradual decline in demand for new cars. The price increase resulting from these changes should be no more than 30%. Porsche showed a concept car incorporating these ideas at the 1973 Frankfurt Motor Show in the shape of a compact two-door four-seater saloon with a 2.5 litre engine.

The LLC is worth considering again in the 1990s as many of the materials choices currently considered are also likely to lead to a significant extension of a car's service life (Nieuwenhuis & Wells, 1994). On the other hand, many of the advances suggested by Porsche have since been adopted in mainstream production cars. As a result, lifespans of cars are already lengthening.

Volvo LCP 2 (1979/1983)

The Volvo LCP 2 was another concept car prompted by the energy 'crisis' of the 1970s. The project was led by Rolf Mellde, the man behind a number of Saab designs, and its main brief was an assessment of the demands to which the automotive industry and its products would be subject by the beginning of the 21st century (Lindh, 1984). Particular attention was to be paid to the materials, design techniques and production methods which might be in use at that time. The weight of the cars was not to exceed 700 kg and maximum fuel consumption was to be 0.04 l/km. Considerable use was made of aluminium, magnesium and various plastics. The aluminium load-bearing structure weighs only 116 kg (Giannini, 1983: 62). Two prototypes were built using plastic body structures.

With the LCP 2 (or LCP 2000), Volvo moved beyond merely developing a highly fuel-efficient prototype. The Light Component Project 2000 was one of the first to take a life-cycle approach. As Volvo put it at the time, it was based on the principle of 'minimisation of lifetime energy' (Janicki, 1992), taking into account the total energy consumption of raw materials, production, use and recovery of the materials. At the same time, it still had to satisfy consumer demand by not moving too far away from existing vehicles in the market. Preliminary figures indicated that the total energy requirement of the LCP was around half that of a conventional car of equivalent size.

The LCP 2000 was developed with IAD (an independent engineering consultancy in the UK) and took an estate design concept in the Volvo tradition. It featured two forward-facing seats in the front and two rearward-facing seats in the rear. It was powered by a turbocharged direct-injection three-cylinder diesel engine. LCP 2000 prototypes were built to allow a realistic test programme. Lessons from the programme are said to have been used in subsequent Volvo production cars.

Renault EVE (1979)/Peugeot VERA (1980–83)

The Economy Vehicle Elements (EVE) was developed by Renault purely as a study in fuel economy in the wake of the 1973/4 fuel crisis. It was carried out in co-operation with the French government energy-saving department. The term 'elements' refers to the fact that only a limited number of potential fuel-saving measures were considered; namely aerodynamics and engine/transmission management. The aerodynamics were to be realistic, allowing proper passenger accommodation, luggage capacity, comfort, safety and performance. This was achieved partly by using the basic structure of the R18. None the less, Renault achieved a Cd of 0.239. Electronics were used to optimise engine and transmission performance. The car combined a top speed of 170 kph with remarkable fuel economy. At around the same time, Peugeot developed the VERA (Vehicule Econome de Recherche Applique) series of concept cars. Unlike the Renault EVE, the VERA cars, though largely based on the 305, also focused on weight reduction by adding mainly plastic components. The cars also used direct-injection diesel engines.

MG EX-E (1985)

At the time of the MG EX-E, Rover was involved in the development of an aluminium-intensive vehicle programme in conjunction with ALCAN. This was largely a weight-reduction exercise and various mainstream Rover models of the time, such as the Metro, were replicated in aluminium. Although the EX-E looks like yet another styling exercise, it did in fact incorporate various elements from the aluminium programme and as such is significant in this context. The EX-E used a bonded aluminium frame with injection-moulded polypropylene outer panels. Unfortunately for Rover, after the BMW takeover, all rights to the technology passed to ALCAN — so preventing Rover from pursuing further development of this particular project.

Renault Vesta 2 (1987)

The VESTA programme for a fuel-efficient vehicle was initiated by the French government, and Renault built three generations of vehicle to try and meet its goal of a fuel economy of 33 km/l. Renault planned to achieve this through improvements in rolling resistance, weight reduction, aerodynamics and powertrain. In all, seven prototypes were built; however, only the last generation, the VESTA 2, of which only one was built, achieved the required fuel consumption figure. The car was built on a R5 platform, but featured a completely new plastic body of much better aerodynamic performance. Despite its length of only 3.53 m, it achieved a Cd of 0.22. In addition, smaller 28-cm wheels were used with electronically controlled ride height adjustment. The car also had an electrically controlled cooling system. Beside the plastic body, further weight reduction was achieved by using very thin, fixed glass for windows, as well as lightweight seats and extensive use of magnesium in powertrain components.

Greenpeace revived interest in this car when it organised a presentation at the Frankfurt Motor Show in 1995, where it put this car forward as the green car of the future — and one that could be built today, much to the embarrassment of its makers.

Matra M25 (1989)

Although the Matra M25 was built as an out-and-out performance machine, much of this was achieved through lightweighting. The car was conceived to celebrate Matra's 25th anniversary (Bellu, 1989: 42). It was a two-seater, measuring 3.47 m and weighing only 675 kg. Fitted with a 1.8-litre Renault turbo engine, the car reached 100 kph in 5 s and a top speed in excess of 250 kph. The low weight combined with high rigidity was achieved with a monocoque body constructed in carbon fibre and kevlar, with an aluminium honeycomb-reinforced floorpan. Considerable attention was also paid to both passive and active safety (Bellu, 1989: 48). Three prototypes of this 'sensible supercar' were built.

Audi Avus and Spyder (1991)

The Audi Avus and Spyder were primarily designed to promote Audi's development (in conjunction with ALCOA) of the aluminium spaceframe technology later used in the A8. The Avus was clearly inspired by Auto Union racing cars of the 1930s and used some styling elements from these cars. The most striking feature of the Avus was its highly polished aluminium bodywork, which was clearly designed to attract attention to the use of this material (CD&T, 1991a). Soon after the Avus, Audi did in fact show a version of the A8 in polished aluminium. The Spyder was conceived as an alternative way of promoting the lightweighting properties of the Audi–ALCOA spaceframe concept (CD&T, 1991b: 20). For a while Audi had the intention of putting this car into production, thus showing the benefits of the technology in both a top of the range luxury car, the A8, and a high-performance sports car.

Volkswagen Chico (1991)

The Chico was one of VW's contributions to the urban electric car debate. However, it differs from most small EVs in being a petrol–electric hybrid. It is intended as part of VW's view of a 'park and ride' style concept whereby people wishing to enter the city would transfer to the Chico. Despite being 10 cm longer than a Mini, it is only a 2 + 2 seater; however, it is state of the art in terms of crash resistance. The Chico is conventional in that it is made as a sheet-steel monocoque, which reflects VW's views of the future. However, it does have some refinements. The front crash tubes will absorb impacts at up to 24 kph and are replaceable if necessary, while the longitudinal members are seam-welded to add strength. The car weighs 785 kg (CD&T, 1991c: 25).

BMW E1 (1991)

The E1 was BMW's first purpose-designed electric vehicle, after several years of experiments with electric-powered versions of the 3-Series. The body/chassis structure was designed in the context of low weight in order to compensate for the weight of the battery pack, as well as low volumes in line with the expected demand for such a vehicle in the wake of California's ZEV legislation. The E1 has an aluminium spaceframe type chassis and plastic outer skin. The skin panels could be manufactured from materials recycled from BMW's mainstream models. The E1 has a total weight of 900 kg, of which 200 kg represents the batteries.

Citroën Citela (1991)

The Citela (CITy ELectric Automobile) was conceived as an urban electric vehicle. It features a lower power module onto which a removable plastic body is fitted. This could be changed to fit differently styled bodies for different applications. The vehicle was designed within the context of PSA's involvement in a number of French government-sponsored electric vehicle programmes and, if these prove successful, PSA has announced its intention to put the Citela into production, possibly as early as the late 1990s. The lightweight composite structure was chosen to compensate for the high battery weight and is an increasingly common choice for small electric vehicles. There is an additional feature to the Citela and this is its design for a long service life. PSA makes a point of stating that the motor will last a million kilometres and that battery life is an exceptional 10 years under normal use.

Volvo ECC (1993)

The Environmental Concept Car (ECC) was Volvo's proposal for a Volvo for the future, based on its principle of weight reduction without downsizing. This was one of the most practical 'foresight' vehicles produced to date, and it used a hybrid powertrain involving a gas turbine and an electric motor. The turbine is used to generate electricity via a series-hybrid set-up, allowing the car full electric drive at all times. Though built on a Volvo 850 platform, the ECC allows weight-saving through an aluminium body which combines a low Cd of 0.23 with traditional Volvo styling cues and the higher levels of comfort the consumer has become used to. Despite its dual powertrain and battery pack, the car is slightly lighter than the 850.

Pininfarina Ethos (1992–94)

Pininfarina, a famous Italian coachbuilding and design firm, developed their Ethos series to show that environmental thinking in car design leads not only to cleaner cars, but also to better cars that are more fun to drive than most modern overweight and gadget-laden devices. The first Ethos was an open sports car (Frère, 1992), the second virtually a coupé version of the first, while the third was a small saloon. All three feature a spaceframe type chassis welded up from aluminium extrusions, designed in co-operation with the Danish division of Hydro Aluminium of Norway. This is clothed in a body made of Noryl GTX^{TM} thermoplastic panels supplied by General Electric Plastics (Frère, 1992: 17). The Ethos cars are designed with ease of manufacture and relatively low cost in mind, as Frère (1992: 17) points out. The sports cars are fitted with a small Orbital two-stroke engine, and Ethos 1 weighs in at 700 kg. Of this, 100 kg is aluminium: 70 kg for the frame and 30 kg for the engine cover and alloy wheels.

Fiat Downtown (1993) and Zic (1994)

The Downtown and Zic are both small electric vehicles designed for use in an urban environment. Both incorporate Fiat's unique expertise in small vehicle design. The Downtown was conceived to 'provide the motorist with the mobility he has lost in city centres'. It should also be environmentally friendly by means of limited energy needs and maximum recyclability. The three-seater body is 2.5 m long and it weighs 700 kg, including batteries. Recyclability is achieved through intensive use of aluminium and recyclable plastics.

Zic takes the small urban EV concept one step further. It is part of a government-sponsored project, via the CNR (Centro Nazionale di Ricerche; Italian national research centre), into special materials for advanced technology. It is a four-seater with a total length of 3.24 m, but a kerb weight of only 860 kg, including batteries (Calliano, 1995: 57). Zic uses an aluminium spaceframe structure, with extrusions welded and bonded to joining nodes. This is covered with sandwich-structure polymer composite panels (made using RTM), including a composite polymer floorpan. Seats are in magnesium alloy, while innovative glass technology gives a 40% lighter windscreen without loss of rigidity. The total body/frame structure consists of only 32 components, compared to the more usual 150 components on cars of this size in current production. This has contributed to a 30% reduction in weight and a 20% increase in structural rigidity, as well as enabling a reduced time between design and manufacture.

PSA Tulip (1995)

PSA describe the Tulip (Transport Urbain Libre Individuel et Public) as 'A vehicle at the cutting edge of automotive technology.' It is a small battery electric vehicle for urban use, and forms part of PSA's integrated system of recharging stations and vehicles to provide clean personal public transport in an urban environment. However, PSA have opted for a radical design with futuristic styling, and a body construction concept derived from PSA's participation in the CARMAT (new materials) and RECAP (recycling) programmes.

The Tulip body structure is the first product from the Peugeot-Citroën Research Centre's CBC (composite body concept) project, which aims to reduce the number of body parts, the number of different materials used and the vehicle production lead time. The vehicle has a body that consists of only five major components made of rigid cores of polyurethane foam covered with polyester-resin injected fibreglass-reinforced panels (RIM), which are bonded together in the assembly process. These parts are structural and form both the interior and exterior of the car. Seats, dashboard, etc., are all integrated with the structure. Colour is added in the moulding process, thereby obviating the need for a paint shop, and the plastic material is also resistant to dents and minor collisions, which leave the body largely undamaged. The lower body parts are made softer to protect pedestrians from injury. In addition, the materials are chosen with ease of recycling in mind and the car is 95% recyclable. The Tulip is only 2.2 m in length (a Mini is 3.05 m) (Figure 6.1).

NedCar ACCESS (1996)

The ACCESS (Aluminium-based Concept of a CO_2 Emissions Saving Sub-compact car) was developed by the semi-autonomous product development division of NedCar, the Dutch joint venture company that makes the Mitsubishi Carisma and the Volvo S40/V40. It was first shown at the Geneva Show in 1996. What is interesting about ACCESS is the fact that it is a realistic and practical running prototype incorporating leading-edge technologies. It represents a vision of a car of the near future, but it departs quite radically from today's technologies, particularly in terms of body structure.

The ACCESS body consists of an aluminium semi-spaceframe involving a mix of extrusions and sheet metal. This is clad in panels of two different materials. The horizontal panels are made of Hylite^{TM}, an aluminium–plastic–aluminium sandwich material, while the vertical panels are made of a thermoplastic material, Stapron-N. Suspension, too, is made largely of aluminium sections,

Figure 6.1 The Peugeot Tulip Urban EV Concept

including extruded front wishbones. The four-cylinder engine developed specifically for the car is half the weight of a conventional engine and runs on the lean-burn principle. It is mated to an automated manual gearbox and relies strongly on electronic control systems.

Concept 2096 (1996)

Concept 2096 is a very futuristic concept of what cars may look like a hundred years from now (Figure 6.2). It was commissioned by the UK Society of Motor Manufacturers and Traders (SMMT) for display at the 1996 British Motor Show, which it organises. It was intended as a celebration of the first centenary of the British motor industry. The work was carried out by Coventry University's internationally acclaimed industrial design department and took the form of a full-size mock-up.

The shape of the 'vehicle', which has been described as resembling a slug, stems from the technologies assumed to have been developed by 2096 and incorporated in the concept. These include:

- a driverless navigation system (no need for steering wheels, or driver skills!)
- the ability to combine with other vehicles as part of high-speed convoys
- 'slug drive' motion using 'muscular genetic rubber' and rendering the wheel obsolete
- powered by fuel cells
- the absence of windows

Figure 6.2 Concept 2096: What We Will Be Driving 100 Years from Now?

In that it incorporates technologies as yet unknown, Concept 2096 is almost a return to the fanciful creations of the 'dream car' age. However, some of these ideas are likely to be realised well within the next hundred years.

6.5   US CONCEPT CARS

In a sense, the concept car is almost a US invention. The Big 3, especially GM, produced several 'dream cars' in the 1950s to stimulate demand and help create the great American car culture. These cars incorporated aerospace styling elements and carried names such as Golden Rocket and Firebird I, II and III. Some of these incorporated experiments with alternative body materials. Of the Firebirds, for example, number I featured an aluminium body, II a titanium body and III a GRP body. Other genuine innovations were also found on these cars.

At the same time, the US started experimenting with alternative power sources such as electricity. GM's Electrovair was an electric version of the Chevrolet Corvair. The most adventurous at this time was Chrysler, which pursued gas-turbine technology and built several working prototypes. A number of interesting concept cars were also shown by Ford and GM in the 1960s. We will discuss examples from each of the Big 3.

Chrysler Gas Turbine Car (1963)

Chrysler's experiments with gas turbines date back to before World War II.

Throughout the 1950s, Chrysler engineers attempted to adapt their expertise in this area to cars. In 1956, a Plymouth saloon with gas turbine successfully completed a drive from New York to Los Angeles in four days. Several other such endurance tests followed. However, the first purpose-designed gas-turbine prototype was built in 1963. In fact, 50 of these cars were built and loaned to 200 selected motorists for a three-month period in order to conduct field trials. The cars looked slightly unconventional, although overall styling conformed to the fashion of the times, belying the radical technology within.

Chevrolet Astro 1 (1967)

The Astro 1 was almost a harking back to the dream cars of the 1950s. It was conceived as a two-seater sports car whose main feature was an unique entry and exit system. The rear half of the body hinged upwards, taking the seats up with it. The occupants then sat in the seats upon which the unit was lowered again and the passengers seated inside the car. Astro 1 used a rear-mounted Corvair engine, though fitted with new overhead camshaft cylinder heads. The car was designed with a minimal frontal area and optimised aerodynamics. The body was made of GRP, although it incorporated an aluminium roll-bar and windscreen surround as structural reinforcements. The car also had magnesium wheels and lacked a conventional steering wheel. Instead, two handles were used to control the car, while flaps were used to aid high-speed braking.

Ford Aurora (1964)

The Ford Aurora was first shown at the New York World Fair. It was essentially a large estate car, as Ford felt that this type of vehicle was growing in popularity. The car had a luxurious rear compartment and the front passenger seat swivelled round to face backward. TV and an elaborate sound system were fitted, together with air conditioning and a refrigerator. Behind this compartment was a bookcase and a glass partition, which divided it from the children's compartment at the rear of the vehicle. The Aurora also incorporated large areas of polarised glass for temperature control and a navigation system. Like other concept cars of the time, the car did without a steering wheel, using instead a small hand grip linked to a power steering system. In a sense, the Aurora presaged the lifestyle concept of the people-carrier or MPV, and is therefore significant.

1970s and 1980s

The energy crises of the 1970s shocked Detroit and most research effort was diverted into so-called 'downsizing', i.e. making cars that were smaller. The AMC Pacer showed that this was not always accompanied by weight reduction, which illustrated the problems for US automakers at this time. Few resources were left for adventurous concept cars until much later. However, in recent years the US auto industry has shown a number of innovative concept cars, some of which we will discuss here.

General Motors Impact (1990)

The Impact was designed as a showcase for the various high-tech divisions within GM worldwide, while adding credibility to GM's electric vehicle programme set up to meet Californian ZEV legislation from 1998. The car was shown in prototype form and was originally intended for production from 1994. However, GM's parlous financial situation caused this to be postponed. A production version, the GM EV1, was launched in December 1996 in selected markets. GM is using its existing Saturn franchised dealership network to distribute the car. Consumers have not been given the option to purchase the car; instead, GM has made it available on lease terms.

Impact is an electric sports car and in tests it out-accelerated a Nissan 300 ZX and a Mazda MX5 to 60 mph. Apart from innovative electric vehicle technology, the car features a very lightweight body structure made up largely of composite materials. Impact boasts impressive aerodynamics, with a Cd of only 0.19 (Brambilla, 1990: 74). The fibreglass-reinforced plastic monocoque gives the car an overall weight of 997 kg, of which 394 kg is accounted for by the batteries. Interestingly enough, this composite monocoque was intended for production, although the EV1 uses an aluminium body. Among the advantages of the various technologies used in Impact was its expected lifespan of at least 30 years. It was described by senior GM personnel as: 'A car to pass on to your children'.

GM Micro (1990)

The Micro was a genuinely small car which was used primarily to display one of GM's first prototype Orbital two-stroke engines. These engines tend to be around half to two-thirds the weight and size of conventional internal combustion engines for a given power output, making them particularly suitable for small lightweight cars.

Ford Contour (1991)

The Ford Contour showcar presented at the 1991 Detroit Show incorporated some genuine innovations, making it one of the most interesting US concept cars in recent years. The car incorporates a novel T-drive powertrain, which involves a transversely mounted straight-eight engine with a central power take-off from the middle of the crankshaft to an in-line mounted gearbox. This unique packaging allows a much more compact FWD layout than existing designs and is likely to be used in future production cars. It also allows a relatively uncomplicated 4 × 4 arrangement — also featured, this is usually a problem with transverse engines. The T-drive also allows the use of a weight-saving transverse leaf front spring, which combines an upper link and anti-roll-bar function.

The Contour features an aluminium spaceframe in which interlocking connectors are used to link the different subassemblies through bonding. Ford estimated that this spaceframe approach allowed a 60% reduction in tooling costs. The car included a number of components redesigned for weight-saving. The cooling air was ducted away from the radiator and fed out through the wheel arches in order to reduce the aerodynamic disturbance caused by the wheels. Some elements from this car can be expected in Ford production cars by the late 1990s.

GM Ultralight (1992)

The Ultralight is one of the most interesting concept cars shown in recent years. It was a radically styled four-seater with the potential of returning 2.9 l/100 kg with sporty performance. It uses a small two-stroke engine, advanced aerodynamics and a weight of only 635 kg. The body consists of a carbon-fibre monocoque with enough rigidity to allow large gullwing doors and an absence of B-pillars. The 1.5-litre three-cylinder engine is mounted transversely behind the rear seats and feeds into a Saturn four-speed automatic transmission. This is all mounted on a 'power pod' subframe, allowing different power options to be installed relatively easily, such as a hybrid powertrain. The car has a Cd of only 0.192, while self-levelling air suspension is used to compensate for the large difference between laden and unladen weight typical of ultralight cars; passengers can account for some 40% of gross weight (Stoddart, 1992: 24). However, the material costs for the carbon fibre alone are put at US$13 000, although crash safety is said to be impressive.

Ford Connecta (1992)

The Ford Connecta was an electric car with bodywork by Ford's Ghia design house in Italy. It was essentially a small MPV with four fixed seats and two additional folding occasional seats behind. The vehicle was built on the floorpan of a Ford Ecostar van, based on the European Escort. Connecta had a composite body and used a high-temperature sodium–sulphur battery pack.

Plymouth Prowler (1993)

The Prowler is essentially an updated pastiche of the classic American 'hot-rod' (Figure 6.3). The concept car used components from Chrysler's LH model to keep costs down, although its body structure combines aluminium and SMC components. Chrysler has put a version of the Prowler into low-volume production and the alternative body materials have been retained in the production version. The body engineering is carried out by the US division of the UK-based Mayflower company.

Figure 6.3   The Aluminium-Intensive Plymouth Prowler Concept Car, Now a Production Reality

Jeep Ecco (1993)

Another Chrysler concept car was the Jeep Ecco, a small 4 × 4 sports utility vehicle with a plastic body, two-stroke engine and six-speed gearbox. The vehicle also features a full-length canvas top. The body/chassis structure combines plastics and aluminium parts, and is said to be totally recyclable.

Chrysler Atlantic (1995)

The Atlantic was a retro-styled concept car inspired by the Bugatti 57SC Atlantic of the 1930s. It has been described as a piece of mobile sculpture in that the primary consideration was its aesthetic appeal. Like the Bugatti, it was powered by a straight-eight, produced by fitting two four-cylinder engines end on. It was based on a Viper chassis, but is unlikely ever to reach production. In a sense it spells a return to America's 1950s dream cars, designed primarily to celebrate the joys of motoring and the motor car.

Chrysler China Concept Car (1996)

To quote from Chrysler's own press release for this car, 'The China Concept Vehicle (CCV) was developed as a people's car for an expanding automotive segment, positioned between a motorcycle and a traditional entry-level car or truck.'

This little car bears a remarkable resemblance to the Citroën 2CV, although it is technically different in many respects. The CCV was designed to be attractive, durable and comfortable for five people and their cargo. It is designed to travel over rough roads and to require a minimum of maintenance. The body is made from colour-moulded recyclable composite material on a steel chassis. It is powered by a small 800 cc two-cylinder air-cooled engine (like the 2CV), which should be capable of delivering around 50 mpg (4.7 litres/100 km). Overall weight is a very respectable 544 kg. Chrysler points out that the CCV has only 1100 parts, compared to the more usual 4000 + of a contemporary car. Like the 2CV, it has a full-length fabric sunroof and three-stud wheel fixing.

6.6   JAPANESE CONCEPT CARS

For a time during the later 1980s and early 1990s, the Japanese producers showed vast numbers of concept cars at the annual Tokyo Motor Show. Many of these were in fact of little value and did not change our thinking on the future of the motor car. However, there were a few interesting vehicles which deserve further attention. Japanese concept cars are on the whole more interesting from the point of view of powertrain technology than body technology. This reflects the priorities of Japanese car makers, who tend to take steel very much for granted as the most suitable body material for cars.

Japanese concept cars started in the 1960s and were used largely to gauge the Japanese public's response to new styling approaches rather than as test-beds for novel technologies. In addition, under the influence of the US at that time, many incorporated safety features. Many cars took the form of dream cars at this time, such as the Toyota EX-III of 1969, or the Mazda RX-500. A more interesting car was Toyota's RV-2 of 1972, which was a modular multi-purpose vehicle. In the same year, Toyota also showed its ESV experimental safety car.

The Honda HP-X shown at the 1984 Turin Show used composite materials such as kevlar and carbon fibre in its bodywork, although this was more a Pininfarina product than a Honda. Japanese innovation during the 1980s was more in powertrain and chassis technology than in bodywork, and it is these areas that enjoy the greater attention in their concept cars. Some of the more interesting concept cars of this period include the Mazda MX-81 of 1982 (produced in co-operation with Bertone), the Nissan NX-21 of 1985 and the Toyota FXV of 1985.

Nissan CUE-X (1987)

One of the most complete concept cars of the mid-1980s to emanate from Japan was the Nissan CUE-X. It illustrates the Japanese preference for chassis and powertrain technology in concept cars. The car featured a highly sophisticated V6 engine, the VG30. This 2960 cc 24-valve engine produced more than 300 bhp with the aid of two ceramic turbochargers with intercooling, direct ignition and variable intake manifolding, which together with valve timing and throttle were electronically controlled.

The car had 'intelligent' four-wheel drive using an electronically controlled continuously variable front–rear torque split. The car also had electronically controlled air suspension which could work as active suspension. The CUE-X further introduced Nissan's HICAS four-wheel steering system. The body styling was relatively conventional, although quite pleasing and appropriate for the luxury segment for which it was intended.

Toyota AXV-IV (1991)

Toyota has produced a concept commuter car made from lightweight materials. This small two-seater incorporated aluminium, magnesium and carbon-fibre-reinforced composite plastics. The car was 340 cm long, weighed only 450 kg, and was powered by a 64 bhp, 804 cc two-cylinder two-stroke engine employing five valves per cylinder. The engine itself weighed only 83 kg.

Mitsubishi HSR-III (1991)

The MMC HSR-III (High Speed Research) was a sports car concept vehicle embodying a number of Mitsubishi Motor Company (MMC) technologies. The car was intended to be optimised for energy saving and for this reason MMC described it as having 'environment friendly performance'. The plastic two-seater body was combined with lightweight steel, and aluminium chassis and powertrain components. The car was also optimised aerodynamically. It used a very small 1.6 litre V6 engine with 24 valves producing 180 ps at 8500 rpm. The car used MMC's electronically controlled four-wheel steering, active suspension and automatic 4 × 4 transmission. The 'recyclable' plastic body contributed to a weight of 1100 kg.

Nissan CQ-X and AQ-X (1995)

These cars embody Nissan's vision of weight reduction without downsizing. This has been achieved by optimising the interior space to the dimensions of an executive car whilst still being compact overall, with a total length of 432 cm. The reasons for taking this approach were environmental: the objectives were to 'reduce pollution and raise fuel efficiency by weight reduction' (Nissan press release).

Nissan described the car as being 40% lighter than typical D segment cars. This has been achieved by extensive use of aluminium in the bodyshell, and overall weight was only 850 kg. The body used hydroformed aluminium extrusions, while a new lightweight sound-absorbing material was also used. This weight reduction has allowed downsizing of other components, particularly drivetrain where direct-injection petrol and diesel engines are combined with

Figure 6.4 Mitsubishi's HSR-V Concept Car

electronically controlled, continuously variable transmission (CVT). The car was also fitted with a range of electronic systems such as satellite navigation, night vision, etc.

Mitsubishi HSR-V (1995)

The HSR-V was the latest in a series of HSR concept cars from Mitsubishi Motor Co. (MMC). In this latest incarnation the term stands for 'Harmonic Science Research', and is used to showcase safety and environmental systems. It is a small two-seater mid-engined sports car with an overall length of 390 cm; it weighs 980 kg and is fitted with a direct-injection petrol engine (Figure 6.4). The plastic body is fitted to a carbon-fibre-reinforced aluminium chassis, which MMC claims greatly enhances crash performance. Great attention has been paid to driver ergonomics, and both the seat and the vehicle height can be adjusted over a 15 cm range to adapt to high-speed or off-road modes, but also to ensure appropriate fields of vision for urban and high-speed use. Driver information systems are also a major feature of the car.

Daihatsu Town Cube (1995)

The Daihatsu Town Cube is in fact a small commercial vehicle. However, it illustrates the way in which legislation can determine vehicle design. The Town Cube is optimised for the Japanese maximum dimensions for the 'Kei' class or 'microcars'. It is a large mobile box that fits exactly into the 3300 mm × 2000 mm × 1400 mm dimensions specified by law.

6.7   CONCLUSIONS: THE MESSAGE FROM THE INDUSTRY

These concept cars may never reach production, or even prototype stage in some cases, but they do show a number of things:

- The industry is preparing for radical change.
- The industry is expecting the main forces for change to be reduced fuel consumption and reduced emissions.
- The industry is seeking the solution to these in alternative body structures and alternative powertrains; notably aluminium or plastic bodies, or combinations thereof, powered by battery-electric or hybrid-electric powertrains with improved internal combustion (e.g. Orbital two-stroke, direct-injection petrol) as a short- to medium-term alternative.

Interviews with key personnel of vehicle producers confirm the preparedness of the industry; however, the only real trigger to start a more radical move in the technological direction indicated by the concept cars would come through legislation. The most likely candidate in Europe is the introduction of some kind of carbon tax or CAFE-type legislation designed to reduce CO_2 emissions and fossil-fuel dependency, while in the US a tightening of CAFE regulations would be the trigger.

Within the motor industry, one of the main deterrents to introducing radical new technologies is the expected lack of market acceptance. We will return to this point in Chapter 7, although the analysis in Chapter 2 is also relevant in this respect in that it shows the current market situation. Concept cars have only limited value as a means of market testing, though in some instances this has been achieved. A recent example occurred when GM showed several possible derivatives of the Corsa model. Press and public reaction to these prompted the introduction of one derivative, the Tigra, which has proven quite successful.

Ironically, the more radical the concept, the less useful it is as a means of gauging market response or of proving production technologies. The major variable here is, of course, the cost per unit of series production, and without this information it is very difficult to arrive at a reasonable forecast of consumer demand. This in turn reinforces the advantages of designs which move away from the all-steel approach, because such designs can be introduced at much lower volumes and overall production levels can be expanded incrementally in line with demand should the product be a success. We will return to some of these points in Chapters 8 and 9.

7 Towards the Environmentally Optimised Vehicle

7.1 INTRODUCTION: TOWARDS HOLISTIC REGULATION

For critics such as Yanarella and Levine (1992), the concept of sustainable development which has come to occupy a prominent role in many state intervention policies (Fukukawa, 1992; Meana, 1992; Schmidheiny, 1992) is a contradiction in terms, and can only represent a transitional state on the way to full sustainability. In the automotive industry, as we have shown, regulation has to date been limited in scope and has failed to achieve the degree of progress desired. This perhaps is also reflected in the wider debate over government regulation and industry in which there appears to be a growing consensus in favour of voluntary agreements and other flexible, informal regulatory arrangements, under a so-called corporate environmentalist regime, rather than the mandating of performance targets (Gouldson, 1996; Seyad et al., 1996; Verheul & Termeer, 1996). Maxwell (1995: 157) illustrates how this approach may prove attractive for industry, government and sometimes also the public at large, although he has to admit that, 'While industry's self-imposed standards by definition exceed current regulations, they may fall short of those that would have been mandated in the absence of corporate environmentalism.'

From material choice, to methods of manufacture and sale, vehicle use and disposal, at all stages important environmental issues are raised. Equally, it is clear that the environmental calculus for the automotive sector must include an account of the infrastructure costs of cars, and the extent to which that infrastructure itself offers a barrier to radical change. The purpose of this chapter, then, is to provide a framework with which to analyse the industry, as a contribution to the wider debate on sustainable transportation (OECD, 1992; Bannister & Button, 1993; Farrington & Ryder, 1993). In so doing, we are supporting not less regulation, but more. This is not to deny the importance of the market; consumers (or markets) will be the final arbiters of success or failure. Our contention is that government does, must and will continue to define the 'shape' of the market in all sorts of ways (the market is never 'free' therefore) — and must continue to use regulation to provide clear market signals to consumers to encourage the adoption of environmentally optimised vehicles.

Thus far, the emphasis of the response from the automotive industry has been on two main themes: refinement of existing technologies and design philosophies (in a way that mirrors the intensification of the work process under lean production); and a more marginal experimentation with more radical alternatives. The EOV proposal is to provide the market conditions for alternative technologies to flourish. Of course, a feature of regulation which we have drawn attention to elsewhere in this book (see Chapter 3) is that it is partial in character, and indeed that regulation in one area may conflict with regulation (or non-regulation) in another. This leads to sub-optimal solutions because vehicles are designed and engineered only to meet specified regulations. Reconciling these conflicts will not necessarily be easy; the intention of the EOV concept is to illustrate how a more holistic approach to regulation may be achieved.

The car affects our environment in a number of ways throughout its lifespan, and these are summarised in Table 7.1. Energy is used for each stage and, potentially, pollution is created at each stage. Attempts to tackle these problems should therefore centre around improving energy efficiency in every area and developing alternative energy sources, both for the production phases and the use phase of cars and trucks. Energy is a useful starting point, or common denominator, to develop a simple proxy of environmental damage.

Table 7.1   Car Life-cycle Environmental Impact

---

1. *Pre-assembly*
   - mineral extraction for raw materials (iron ore, bauxite, oil, etc.); transport of raw materials
   - production of secondary materials (steel, aluminium, plastics, etc.)
   - transport of these materials to assemblers and suppliers
   - production of components and subassemblies
   - transport of components and subassemblies

2. *Assembly*
   - energy use in assembly plant
   - pollution caused in assembly process, particularly paintshop emissions
   - release of waste materials into ground and water and into the recycling system
   - transport of finished vehicles to customer

3. *Use*
   - energy used for driving
   - pollution caused by emissions and waste materials from disposables (batteries, tyres, oil, etc.)
   - land-use requirements (roads, fuel stations, parking facilities, etc.)
   - accident damage to people and environment

4. *Post-use*
   - transport to dismantling site/scrapyard
   - energy used in dismantling/scrapping process
   - pollution caused by dismantling/scrapping process
   - transport of recyclates

The land use requirements lead to the congestion problems, and solutions for this are sought in alternative transport modes. Much attention has focused on moving drivers from their cars into public transport modes. We do not deny the importance of this debate — the EOV cannot of itself solve the problems of congestion after all. However, there is a distinct difference between private and public transport, and alternative private transport modes such as motorbikes, bicycles (or other human-powered vehicles) and walking may be more attractive if proper facilities are provided.

Focusing on emissions only tackles a very small part of the total picture. This has been recognised by many vehicle manufacturers, such as Volvo with its Environmental Protection Strategy or EPS (Volvo, 1991a, b). Although regulation has historically only sought to define incremental and economically achievable improvements in car performance (and in this sense lag behind what is technically possible), regulation also plays a determining part in defining the market. As a result, market acceptance (i.e. by the consumer) probably lags even further behind than regulation in terms of what is technically possible, while the narrowness and prescriptive nature of much contemporary regulation may actually exclude promising new technologies.

It is clear that modern industrialised society can no longer function without some form of motorised private transport, at least in the short to medium term — though this should not be taken as an excuse to accept the demand for mobility as a given. In these circumstances it seems prudent at least to optimise the environmental performance of the car, and it is here that a more holistic approach to regulation as espoused in the EOV has a role.

7.2 INTRODUCING THE EOV CONCEPT

There is no such thing as a 'green' car. The nearest thing would be a wooden land-yacht; made of wood (a renewable resource) and powered directly by wind (a renewable resource). Such a vehicle has certain limitations, however. None the less, we can improve on what we have, while accepting that every car imposes some sort of burden on the environment. At the same time, industrial societies and many developing countries can no longer survive or support their populations without some form of motorised transport.

Despite its centrality in modern industrial economies, the environmental pressures on the automotive industry and its products are mounting. It is clear that existing technology relies very heavily on ultimately finite resources. Mass motorisation also imposes unacceptable environmental burdens on society in the form of air pollution, congestion and road accidents, and it contributes to a growing waste problem.

As the efforts by manufacturers to clean up emissions become increasingly costly for an increasingly limited reward, a different approach is needed. At present, manufacturers are not rewarded in the market for building cars in clean, low-polluting or low-energy-consuming factories — except in so far as environmental improvements in the production process contribute to reduced costs. They are not rewarded either for building cars that last longer than the average, thus reducing the waste burden and the need to make new replacement cars so frequently. At present, they are also not rewarded for reducing the fuel consumption of cars and thus reducing CO_2 emissions.

In fact, the attempts at ever tighter toxic emissions limits can themselves exacerbate the environmental impact in other areas. Catalysed cars were — initially at least – often less fuel efficient. The higher octane versions of unleaded petrol needed for catalyst compatibility contain higher levels of benzene, which is a known carcinogen with no safe level of exposure and, although cleaned up by a catalytic converter, it imposes a serious health burden in the form of evaporative emissions from fuel stations, tankers and car fuel systems.

Similarly, environmental arguments are often used by the industry to support premature scrappage of older cars. It is argued that these are more polluting than new cars, but this is not necessarily true under all circumstances, and can vary considerably from model to model and pollutant to pollutant. Conversely, premature scrapping does increase the waste burden and wastes energy, while creating an overall increase in pollution (Nieuwenhuis & Wells, 1996; Wells & Nieuwenhuis, 1996).

The natural and human environment would benefit more from real attempts to reduce the car's detrimental impact overall than from merely reducing — admittedly harmful — toxic emissions from each new car. A more global approach is therefore needed which allows manufacturers to tackle the detrimental environmental effects of their products in a whole range of different areas, and be rewarded for it in the market and by the legislator. Under such a system, a below-average performance in one parameter could be offset by an exceptional performance in another. This would also accommodate the trade-offs that often have to be made between different technologies; for example, a diesel engine is good on CO, CO_2 and HC, but not so good on PM and NO_x.

For these reasons, we first proposed the environmentally optimised vehicle (EOV) concept as a basis for future legislation on motor vehicles and environment (Nieuwenhuis & Wells, 1994: 16). It was subsequently worked out in more detail in Nieuwenhuis (1995). The EOV concept sets out a whole range of environmental parameters in which a car has to be optimised in order to qualify. However, it allows trade-offs between the different parameters, and allows recognition to be made of efforts in particular areas not at present covered by legislation. It will also often allow a manufacturer to make a real impact on improving the environment for relatively limited resources, thus benefiting society as a whole more than is possible under the present regime. In addition, any standards achieved on a new car would have to be maintained at a reasonable level throughout its life. Regular testing is an important prerequisite to ensure that the entire parc can be environmentally optimised. It is important to realise that this is not a prescriptive approach, it does not seek to mandate how the vehicle manufacturers should reach the performance targets, or which targets are to be given prominence: the framework is explicitly intended to prevent a narrow 'design for regulation' and to allow latitude for innovation. Thus, the failure of regulation discussed in Chapter 3 should not be taken as a reason to abandon such regulation. Neither is there any evidence to support the view that PNGV-type initiatives will result in environmental optimisation. Table 7.2 gives a broad normative definition of an EOV.

In practice, each of these parameters would be weighted, the weight reflecting the current level of scientific knowledge and public concern. To qualify as an EOV, a vehicle would have to achieve a certain minimum score which could be made up of low toxic emissions (lower than the legal requirements), low energy consumption in production, a long product life expectancy and easy recyclability, for example. The choice would be up to the manufacturer. Cars not qualifying as EOVs would incur a tax penalty of some sort.

Some examples of how such a system would work are as follows. If an individual model achieves a better performance in toxic emissions, beyond the legal standards, it would incur, say, ten credit points for each 1% improvement over the legal standard in independent testing. For carbon dioxide emissions we could take fuel consumption as a measure, as this shows a close correlation. We could take the parc average for the market and give credit points either for any improvement over this average or for an improvement over the model's segment average. The former would include half the parc, although the amount of credit

Table 7.2 EOV Definition

---

1. An EOV is a vehicle which, given existing technology and economic constraints, has

   (a) the lowest possible toxic emissions (HC, CO, NO_x, PM)
   (b) the lowest possible fuel consumption
   (c) the lowest possible 'greenhouse' gas emissions (CO_2, CFC, methane, etc.)
   (d) the longest possible life expectancy/durability
   (e) the lowest possible requirement for harmful consumables (tyres, batteries, oil)
   (f) the lowest possible requirement for road and parking space

2. ... and which in its production

   (a) has the lowest possible raw materials requirement
   (b) causes the lowest possible emissions
   (c) has the lowest possible energy requirements
   (d) has the lowest possible VOC emissions and paint requirements
   (e) incorporates as much as possible recycled, recyclable and renewable materials

3. ... and which in its disposal

   (a) produces the lowest possible waste burden
   (b) produces the lowest possible emissions in dismantling and recycling
   (c) requires the least possible energy input in dismantling and recycling
   (d) has the maximum possible degree of recyclability at the lowest possible cost

---

Source: Nieuwenhuis (1995).

would vary. It would clearly favour small cars. The combination of using an index-based approach with a parc-based approach is interesting because it allows for historic differences between distinct markets to be reconciled. Fundamentally, environmental optimisation is not simply a scientific or technical issue, it is also one of social and political choices. The EOV concept therefore exposes many choices which are currently implicitly made. An additional benefit would be that wider debate on these issues would greatly improve consumer knowledge, thereby allowing more informed market behaviour.

Energy tax proposals circulated by the European Commission (CEC, 1996) appear to underplay a parc approach, favouring instead a segment approach based on new car sales. The expectation is that over time the cumulative effect of new car purchases influenced by the carbon tax would lead to overall improvements in the performance of the parc as a whole through the replacement of existing cars. A natural additional policy is to remove 'gross polluters' with scrappage incentives. Initially, the focus on new car sales could be accepted; however, ultimately, a parc approach, favouring light cars, would have to be introduced. In fact, the Commission's approach does acknowledge this:

> The basis for such a measure would be a CO_2 reference standard incorporated into the vehicle type approval procedures. This standard would be related to vehicle mass, cylinder capacity or horsepower. Fiscal incentives would then be given to vehicles the CO_2 emissions of which are below the reference standard. The standard would be lowered in steps under a specified timetable, with a stronger lowering for bigger vehicles.
>
> (CEC, 1996: 11)

It is also interesting to see the Commission accepting the principle of a gradual tightening of measures according to a reference standard or benchmark, a point we will return to below.

Rating life expectancy is more difficult and can only be based on the company's or model's past record, whereby a 20-year historical horizon is not unrealistic. A step in this direction would be to insist that spare parts for any particular model will be made available for at least 20 years after production of the model has ceased, rather than the current 10-year requirement. The measure would be tested against the average durability in the markets over the period. Credits would be given to companies whose products have a median useful life that is longer than the average.

For the production and transport phase, a rating of energy consumed by an assembly plant by number of cars built could be used (Nieuwenhuis, 1994a). It is already the case that some vehicle manufacturers have produced 'environmental' reports which do precisely this. Some of the variables that might be used are illustrated in Table 7.3, which shows the Volkswagen Golf built at Wolfsburg, Germany.

However, it is important to compare like with like, as assembly plants vary in the proportion of work they carry out on site. In the case of the VW Golf, for

Table 7.3 Measures of Environmental Performance at Plant Level: The Case of the VW Golf

Item	Units per Golf produced
Hydrocarbons — paint shop	2.8 kg
Dust	50 g
Consumption of drinking water	5 m^3
Heavy metals in treated water	620 mg
Specific nitrogen content, treated water	40 g
Specific phosphate content, treated water	4 g
Hazardous waste for disposal	19 kg
Hazardous waste recycling	21 kg
Industrial waste for disposal	11 kg
Industrial waste for recycling	19 kg
Energy	4.2 MWh

Source: VW (undated).

example, Wolfsburg is a relatively integrated site remote from the main sources of components and material supply. Some allowance for level of vertical integration should therefore be built in. None the less, the Volkswagen example shows that this must be measurable. Environmental audits and the provision of measures of system performance under schemes such as EMAS (Environmental Management and Accounting Systems) should rapidly improve the information base required to evaluate the environmental manufacturing costs on a per model basis (Clegg, 1996).

For measuring recyclability, a number of parameters could be used. These would include the number of different plastics used, compared with the parc average; the number of different metals used, against the parc average; how much of the car can be recycled; how much can be dismantled in a given time, say, one hour; how much is reused in the company's cars; how much is reused in the company's other products; how much is reused elsewhere. These last few measures would assess the level of downgrading. A final measure could be the percentage of renewable resources used in the car. This would include leather and wood, crop-based sound insulation materials, etc. It is the case that, with new product launches, the vehicle manufacturers have made claims as to the potential their cars have for being recycled. Ford, for example, claims that 85% of the Contour/Mondeo model as sold in the US and Europe during the mid-1990s can be recycled (Ford, undated), against an industry average of 75%.

It is clearly important that any regulatory changes with respect to recycling should be sensitive to the danger of the potential damage caused to the present recycling infrastructure. In the industrialised countries this infrastructure successfully recovers virtually all the metals in a scrapped car. There can be little doubt that recycling must be profitable if investments are going to be made in this area, and that there is a clear need for better recycling practices. On the other hand, recycling as an activity must be considered a last resort (Nieuwenhuis & Wells, 1996).

In order to qualify as an EOV, a model would have to reach an overall rating within the top 25% or 30% of all cars currently sold in the market. This ensures that the target keeps moving. As scientific and technical knowledge progressed, standards could be tightened and weights changed. This would also fit in with the European Commission's progressive tightening of standards under its carbon tax proposals (CEC, 1996). However, in terms of the overall system these would be minor adjustments, rather than a wholesale overhaul or the type of major legislative change that has been a feature of environmental legislation over the past few decades. It would also retain the 'level playing field' that allows a free competitive environment for the industry.

The EOV system could be accused of incrementalism in that it allows for a gradual adjustment of the minimum standards. In this respect it should appeal to the industry, which favours such a gradualist approach. On the other hand, it takes only one radical new design to reach the market — such as the Mercedes-Swatch MCC — to change the definition of 'state-of-the-art', shift the average environmental rating of the top 25% or 30% and prompt a tightening of a number of EOV parameters. In this respect the EOV concept also allows for radical innovation, something the current legislative framework, with its focus on solutions linked to the existing technological paradigm, does not encourage (Wells, 1996b).

7.3   THE EOV AND LIGHTWEIGHT 'HYPERCARS'

Our EOV concept merely formalises in the shape of a proposal for legislation something which a number of observers and commentators have already proposed as a technological approach to car design. These are generally focused around a lightweight car concept. Interestingly, the Mathis VL 333 concept car of 1946, intended for production (and discussed in Chapter 6), is also based on these basic principles, which in practice often lead to lightweighting, improved aerodynamics and other vehicle efficiency measures. The impending EU energy tax and US electric vehicle boom are returning lightweighting to the agenda.

Since the Mathis, much weight has been added, especially in recent years, in the form of electric motors to drive windows, seats, sunroofs, etc. On a BMW 7 Series, adding an electric sunroof adds 17 kg, electric seats adds 10 kg, while air conditioning adds 25 kg (Nieuwenhuis et al., 1992: 42). In fact, most cars have added weight over the past two decades, either from one model generation to the next or within one model generation (Nieuwenhuis et al., 1992: 41–3; Nieuwenhuis & Wells, 1993). Apart from the energy cost of moving this extra weight about, there is a more direct penalty. Mom and van der Vinne (1995: 47–8) point out that in city driving more energy is now used to drive all the additional equipment than to drive the car. However, it is felt that consumers have become used to these comfort items, hence weight reduction has increasingly focused on the largest single component: the body. The customer does not worry about the body weight after all. In practice, the pressure for weight reduction will be felt throughout the vehicle, and hence throughout the automotive industry supply chain, as we discuss further in Chapter 8.

Various attempts have been made since the 1940s to refocus the attention of the industry and the market on lightweight thinking. Work done at the Technical University of Delft in the Netherlands in the 1970s is of interest in this respect. It was noted even at this time (van der Burgt, 1977: 11) that half of all car mileage involves commuting and business use at an average car occupancy rate of 1.2. This increases to 1.7 for shopping, while leisure and holiday use achieve on average 2.2 and 3.5 people per car respectively (van der Burgt, 1977). In other words, for half the car journeys a lightweight two-seater would be sufficient. Van der Burgt's (1977) article, which was written in the immediate aftermath of the political fuel crisis of 1973/4, places this in the context of energy conservation and puts forward a number of proposals for lightweight vehicles. The aim is to develop a car with a kerb weight of only 250 kg (van Kasteren, 1977: 12). This was described as an ultra-light private vehicle.

The theme was picked up some 15 years later by Amory Lovins and his colleagues at the Rocky Mountain Institute (Lovins et al., 1993). They had the advantage of a number of technological developments since the 1970s, making the whole idea more feasible. They developed the 'Hypercar' concept, which inspired US President Clinton and Vice-President Al Gore, and formed the basis for the research carried out under the PNGV programme. Lovins also rejects incrementalism and instead favours a technological 'leapfrogging' approach, involving 'an artful fusion of the best existing technologies' (Mascarin et al., 1995: 56).

The hypercar concept has been very influential in the US and beyond. It involves a five-seater saloon car with dramatically reduced weight, greatly improved drag and frontal area, and a hybrid-electric powertrain. Lovins argues that this will allow the downsizing of virtually all other systems in the car, thus yielding incremental improvements in weight reduction. He recommends carbon-fibre composites for the bodywork, and recognises the fact that this and his hybrid powertrain mean a radical departure from current car-making practice. He advocates the development of a completely new industry to make such cars (in Cronk, 1995), and relies on computer companies and small firms to bring this about. In fact, although much of the necessary technology resides with small firms, experience at Lotus, for example, shows that the resources of large firms are needed to make it happen (Ken Sears, Lotus, pers. comm. 1992).

Although Lovins recognises that these cars will be more durable (Lovins, 1995), he attempts to show (Mascarin et al., 1995) that they can be made at the same cost as present-day cars. In practice, much greater durability would lead to reduced demand (see Nieuwenhuis, 1994c) and ultimately different forms of ownership. In this context, the increased cost of a hypercar is not a problem and may in fact be a way for the manufacturer to achieve profitability at much lower volumes. Lovins may also be somewhat optimistic about the true performance parameters of his cars, although his contribution has been vital in introducing the idea of moving beyond incrementalism into a very conservative industry.

Another approach to lightweight personal vehicles was worked out and presented by Riley, who points out that 'most of the energy used by the automobile is consumed to transport itself. The energy used to move the occupant is almost insignificant by comparison' (Riley, 1994: 36).

Riley adds that most of this weight is unnecessary most of the time in view of vehicle occupancy rates and trip lengths. He argues, therefore, for simple light two-seater or even single-seater vehicles, as 'Using vehicles that fit driving patterns will greatly improve vehicle utilization and significantly reduce the energy intensity of personal transportation' (Riley, 1994: 38).

He adds that by creating a new type of vehicle, which could be positioned conceptually in between motorbikes and cars, a lot of existing car-based legislation can be by-passed, thus rendering such alternative concepts more viable. Riley points out that one of the barriers to lightweight, more environmentally optimised vehicles lies in the existing body of legislation and regulation, which was created around the present automotive paradigm. Rather than trying to shift this paradigm directly, Riley proposes additional categories of vehicles. Dedicated narrow vehicle lanes would allow a greater utilisation of the existing road network and thus reduce congestion (Riley, 1994: 41–3).

In order for the market to follow such a move, consumers have to experience it as a major paradigm shift. Incrementalism is counter-productive as it carries with it the baggage of expectations of the existing paradigm. These vehicles should not be marketed as cars, but as a completely new type of product. Riley (1994: 45–6) proposes a new set of categories of vehicle as follows:

- *Passenger car:* weight >450 kg
- *Commuter car:* weight <450 kg, freeway-capable
- *Narrow-lane vehicle:* commuter car sub-category for one or two occupants
- *Urban/city car:* weight <450 kg, dedicated for inner-city use
- *Sub-car:* extremely lightweight, including powered bicycles, etc.

However, in order to market such new types of vehicles and gain customer acceptance, Riley (1994: 60–1) explains that, 'The vehicle should not step down to accommodate scarcity and conservation. Instead, it must move up to conquer waste and pollution through superior technology ... New vehicles must be fun to own, touch, care for, and operate. Aesthetic appeal and pride of ownership are independent of vehicle size and energy consumption' (as owners of HPVs will testify).

Riley's idea of adding a new kind of vehicle is particularly interesting. We are very much used to thinking in terms of private transport pigeon-holes: bicycle, moped, motorbike, car. These categories are largely legislation-driven, of course; however, they are considerably more fluid than one might think and it is well worth exploring this fluidity further. Riley proposes adding categories of vehicle between car and motorbike. In a sense, a moped can already be regarded as a category between bike and motorbike. In other countries, such as France, we have the *'voiture sans permis'* (VSP). This is essentially a very lightweight car, with often moped or motorbike technology, with a speed-governed powertrain, which can be driven by people without a driving licence, from the age of 14. These cars are also known as *'voiturettes'*, a term which harks back to the cycle-cars of the period 1910–30. These were very light — and often very crude — vehicles, using bicycle or motorbike technology. Modern materials have greatly enhanced the scope for such products.

On the other hand, there have also been developments in non-motorised vehicles. Although the diamond-framed 'safety' bicycle layout, pioneered by Rover about a hundred years ago, has become the norm, more and more deviations from this layout are entering our streets. Recumbent bicycles are rapidly gaining in popularity, with the first mass-produced example launched in the Netherlands in 1996. These non-diamond-frame bikes are often known as human-powered vehicles (HPVs). Some fully fared (i.e. bodied) racing HPVs have achieved speeds of over 110 kph, largely through optimised aerodynamics and power transmission technology.

Another development in this HPV area is what is now increasingly known as a 'velocar', essentially a sophisticated pedal-car for adults, often using the latest materials technologies. The Danish Leitra has been in production for many years now. It is a three-wheeler with optional body-faring, and room for one adult and shopping or a child seat. The Dutch Alleweder ('all weather'), designed by Flevobike, is an all-aluminium monocoque three-wheeler recumbent with McPherson strut front suspension, available as a kit for home construction (Figure 7.1). Cruising speeds in flat terrain of up to 60 kph are not uncommon, aided by a low frontal area and favourable gearing. With an overall weight of around 30 kg (excluding the driver), this vehicle weighs only about twice as much as a modern mountain bike, and in kit form retails at around £1100.

Another approach is represented by the Twike, an aluminium-bodied two-seater from Switzerland. This vehicle is both pedal-powered and battery-electric-powered. Both systems can be used either separately or together, making this a true parallel EV–HPV hybrid. With batteries, the vehicle weighs in at 240 kg, and has a range of 50 km and a top speed of 85 kph. As an HPV, the weight can be reduced by leaving the batteries behind and, although the top speed will be lower, the range is unlimited. However, at 21 000 Swiss Francs it is not cheap.

7.4 MARKET IMPLICATIONS: THE EOV AS A MARKET CONCEPT

A common criticism levied at those proposing new vehicle concepts is that the market will never accept it. There is a widely held assumption that the car buyer actually wants what is being offered today. The only evidence for this is the fact that he/she buys it. However, car buyers of the 1950s also bought bubble cars, and car buyers of the 1920s bought model T Fords, as well as Bugatti type 35s. The range of cars that people have bought over the past hundred years or so is

Figure 7.1   The Alleweder HPV, Designed by Flevobike of the Netherlands

vast. Henry Ford thought that with the Model T he had optimised the car and no further development was needed. If nobody had come up with a better or more attractive product, we would all still be driving Model Ts.

In fact, the expectations of the average car buyer are limited to what is being offered in the market today, and expectations of the future are also based on what is available today. I remember my father telling me the story of walking in town in the 1950s and seeing a Messerschmitt Kabinenroller bubble car. He turned to my mother and said, 'One day we may be able to afford one of these.' Within 10 years he drove a Volvo, and 10 years later we had three cars in the family; something inconceivable only 20 years earlier. It is also worth noting that the vehicle manufacturers, as with producers of other consumer products, find that the 'market' adopts and uses the product in all sorts of unanticipated ways. It is clear, therefore, that any market niche or segment is initially supply-led, and that only the size of that market and the detailed specifications of particular models can be truly said to be market-led.

The EOV concept allows a number of marketing advantages to the vehicle manufacturers. Historically, the vehicle manufacturers have found it difficult to develop brand image around what might be considered as environmental terms. Making 'green' claims at present not only exposes the vehicle manufacturers to counter-claims from environmentalists, it also conflicts with other (non-environmental) claims made (see Prothero, 1994). An EOV concept would make a significant contribution to establishing the basis on which environmental claims and comparisons could be made. Moreover, the EOV concept has clear resonances with the concept of 'good citizenship' frequently espoused by leading companies.

We do not underestimate the size of the task confronting the vehicle manufacturers in the realm of EOV marketing. A glance at any of the popular motoring magazines in any industrialised country shows a continued emphasis on speed, power, individualism and status. It can be remarkably difficult for vehicle manufacturers to shift market perceptions of the brand or, indeed, to control it completely. An illustration of this problem is provided by Volvo, who over many years developed a reputation for occupant safety in their cars. Unfortunately for Volvo, this safety image then developed into a 'staid' and 'boring' image, while the average age of Volvo new car buyers continued to increase. Central to Volvo's marketing strategy in the 1990s has been an attempt to retain the safety reputation, while enhancing the 'excitement' factor. Volvo have been quite successful (Newton & Iddiols, 1993), but only at the cost of considerable advertising expenditure and two new models, the 850 and S40/V40. So even in this relatively limited realignment of their brand image, Volvo experienced difficulties.

Not all the vehicle manufacturers are similarly placed with respect to the degree of conflict between their current brand image and that which might be pursued under an EOV. Brand image is generally regarded as an asset, but under periods of radical change it might also become a liability: consumers may equate the brand with values inappropriate to those embodied in an EOV.

As we have seen, what is available at any one time has changed considerably over time, and there is no evidence that this process has stopped. The values car buyers attach to the cars they buy today are not permanent and may change. It is therefore important that the vehicle manufacturers can also change in their definition of the brand. The EOV concept provides a first starting point towards re-inventing brand values around sustainable themes. Many in the car industry take a short 15- or 20-year horizon, expecting future developments to take a similar course as those of the past 20 years. History shows us that this is unlikely to be true and that, even what we perceive to be a period of stability, with hindsight turns out to have witnessed major change. Even car producers realise that customer demands for features go through phases: fuel efficiency, safety, luxury, etc. Over time, these features become embedded in expectations and values.

There is no reason why environmental optimisation could not be one of these features; we will not know until the customer is offered the choice of such a product. A 'green' product is not only virtuous in itself, it is also likely to be very driveable. Rather than being a hairshirt machine punishing us for past profligacy, a lightweight EOV can be responsive, fast accelerating and generally fun to drive, as Pininfarina's Ethos concept cars indicate (see Chapter 6). This point is illustrated particularly effectively by a minimalist lightweight concept sports car shown by the Coventry School of Art and Design at the 1996 Birmingham Motor Show (Figure 7.2). Besides, out and out driving machines such as the Lotus/

Figure 7.2   The Coventry School of Art and Design Lightweight Sportscar Concept at the 1996 Birmingham Motor Show

Caterham 7, Lotus Elise, or even the Bugatti 35 of the 1920s, have always been light and relatively efficient compared to the family saloon of their day.

It is clearly not impossible for a manufacturer to make a success of a risky new model introduction. It is quite possible that — as in the case of the Mini, the Chrysler and Matra-Renault MPVs or the Range Rover — there is a potential and untapped market for an EOV-type vehicle. However, the risk is considerable. It is interesting, therefore, that a few producers are willing to take the plunge and move in the EOV direction. Mercedes-Benz is preparing a two-pronged attack on an alternative greener car niche: its 'Smart' joint venture car with Swatch, and its new A Class car. The Smart is a very small car, which may well qualify as an EOV. It will be available from 1998. Similarly, Mercedes' new A class car — a very compact and flexible five-seater, launched in 1997 — could well qualify as an EOV.

In a sense, the involvement of Mercedes-Benz is highly significant. This firm has not only the considerable resources needed, as Germany's largest industrial group, its image and prestige are such that its endorsement of even a radical new concept may ensure success. That is, Mercedes is using its powerful brand image as a producer of high-quality and well-engineered large cars to provide the marketing leverage for an entry into much smaller segments of the market. It is interesting to note that Mercedes started an advertising campaign to launch the A-class well in advance of the model actually being available on the market, and concentrated on emphasising the impact safety of the car, in order to reassure customers that 'traditional' Mercedes values are still embedded in this untraditional car.

The strategy is not without risk — not least in terms of damaging the reputation of traditional Mercedes products. Much of the motoring press thinks in limited terms of present and recent past offerings, and has poured scorn on these cars even before being able to drive them, as witnessed by phrases such as 'Mercedes: the A-class Gamble' and 'the steady hand on the Stuttgart tiller may have twitched out of control' (Rendell, 1996: 34). On the other hand, much of the industry is waiting to see whether Mercedes can make a success of these cars. If it does, many more manufacturers will follow, especially if encouraged by EOV-style legislation.

7.5 CONCLUSIONS: THE POSSIBLE SHAPE OF THE EOV

Having set out the basic concept of an EOV, we can now try and put some flesh on the bones of the conceptual skeleton. Putting together the ideas embodied by the concept cars, as outlined in Chapter 6 and by other observers such as Riley, Lovins and the various contributors to Cronk (1995), we can develop a vision of various EOV-type vehicles (Figure 7.3).

In terms of the body structure, it is clear for the various reasons already outlined that the conventional pressed-then-welded steel monocoque body with steel hang-ons is likely to disappear. The manufacturers, through their concept cars, appear to agree with this and some recent production cars underline this further. The production vehicles listed in Table 7.4 are particularly significant in this respect and we could describe them as 'signpost' vehicles.

In order to combine the demands of weight reduction, longevity (and the possible need for 'retrofitability'), occupant safety and profitability at lower volumes, some sort of 'hybrid' materials solution seems most likely. This suggests some sort of structural three-dimensional chassis, spaceframe or semi-spaceframe in steel or aluminium, clad in panels made out of steel, aluminium or composite. In the longer term, composite plastic or — especially — thermoplastics seem the most likely external cladding media. Thermoplastics are very easy to recycle, especially when compared with thermoset materials such as GRP or SMC; they are also remarkably impact resistant, making damaged panels after minor collisions a thing of the past.

Crash safety performance will be retained or even enhanced by using sacrificial collapsible structures, most likely in steel or aluminium. The latter has proven particularly suitable for this in tests carried out by, among others, Audi and Lotus. Steel has a significant 'first-comer' advantage over aluminium and should not be dismissed as a structural frame material for the future. In relation to aluminium, steel is also plentiful, cheap and forgiving in a manufacturing environment. Both materials will come to be regarded as too vulnerable to denting and too expensive to repair to be used as a cladding (or exterior panel) material. The ULSAB project, outlined in Chapter 4, therefore concentrated on

Figure 7.3   Three EOV Concepts Developed by the Authors

Table 7.4   EOV Signpost Vehicles

Producer	Year	Structure
Europe		
Matra-Renault Espace	1984	Composite over steel frame
Audi A8	1994	Aluminium over aluminium semi-spaceframe
Renault Sport Spyder	1994	Composite panels over welded aluminium chassis
Lotus Elise	1995	Composite over bonded extruded aluminium chassis
MCC Smart	1998	Composite modules over steel
Fiat Multipla	1998	Steel spaceframe with steel panels
Mercedes A class	1997	Double floor design optimises space and safety
US		
Pontiac Fiero	1984	Composite over steel frame
Saturn range	1990	Steel monocoque with composite panels
Plymouth Prowler	1996	Aluminium chassis and panels

the internal structure and adopted many of the characteristics of the semi-spaceframe approach, indicating the key role steel could play in this scenario of a future EOV.

In terms of powertrain, we will be moving away from the petrol–diesel monoculture. By adopting a range of different powertrain technologies, we spread the risk in case one fuel runs out or becomes environmentally unacceptable. Anyway, as Riley (1994: xvi) points out: 'Even if economically recoverable petroleum reserves are three times greater than today's known reserves, we will probably have to abandon oil as a primary energy source for transportation by the year 2020.' Nevertheless, more efficient petrol and diesel engines are likely to dominate for the foreseeable future. Alongside these, we will see an increase in electric traction — either full battery electric for urban use (especially communal personal transport) or hybrid for mixed use.

The advantage of electric power is that it is not an energy source in itself, merely a means of storing energy. This means that electricity gives us a range of generating options, many of which will be available over the long term and some of which — such as solar or wind power — are sustainable. It is remarkable that this issue is not addressed in either the PNGV or the EU 'Car of Tomorrow', despite the environmental benefits of household-scale electricity generation using a combination of wind and solar sources. A widespread adoption of electric-powered vehicles — despite their inherent inefficiencies — therefore gives us some security for the longer term. In this context the fuel cell is also of great interest. In the meantime, various alternative fuels are likely to continue to increase in popularity, especially for urban buses and commercial vehicles.

An EOV is therefore likely to have a steel or aluminium chassis with cladding in composites or thermoplastics. Over time, structural plastic composites will come to be more widely used. It should be emphasised that the EOV concept is not prescriptive, so EOVs may also take different forms. From the examples outlined above, we saw that new technologies are appearing in vehicles ranging from MPVs (Espace) and luxury cars (A8) to lightweight sports cars (Elise) and urban runabouts (Smart). In fact, more varied forms are likely to appear, such as those proposed by Riley (1994), which fit between motorbikes and small cars, or sophisticated human-powered vehicles, using high-technology lightweight materials. Cars may also become more modular, providing a range of different body modules to choose from, as in the Mercedes Vario concept (estate, coupé, pick-up, convertible, etc.), or different powertrain modules, such as in the Opel Twin concept car (internal combustion or electric).

The EOV will also incorporate the much higher electronics content forecast for the industry (Lee, 1994: 98), although this is likely to be focused more towards technology-enhancing systems than towards more gadgets. In this respect, electronic control of engine and transmission will increase, in addition to active suspension, anti-noise, night vision and driver information systems. The greatly reduced weight will increase the difference between unladen and laden weight, as people will not shrink with the car. A 500 kg hyperlight car with five people on board could easily double in weight. Conventional suspension engineering has problems with this, but active ride (as developed by Mitsubishi and Lotus, for example) could have a major role in optimising ride, road-holding and handling characteristics of an EOV. Similarly, anti-noise (developed by Lotus) would allow the removal of most, if not all, of the heavy sound-deadening materials currently found in cars. Electronics tend to offer a weight-effective alternative for many mechanical systems, and solutions such as electronic/electric steering systems are also under development for this reason.

Having provided a brief outline of our vision of an EOV of the future, we must stress that it could take another 20–30 years before these changes make a real impact. The signpost vehicles listed above are all niche products. However, an increasing number of such alternative-technology niche products in actual production adds up to a significant and steadily growing part of the market, while draconian energy consumption legislation – such as a carbon tax – could accelerate these trends. The 'strategic niche' concept developed by Hoogma et al. (1996) is of particular relevance in this context and may be adapted to promote EOV initiatives via local market testing. It provides a model for identifying innovative transport initiatives and hence nursing them and, if successful, rolling them out into the wider market and industrial context. We will try and assess in the next chapter how the changes embodied in the EOV would affect the car-making industry, and examine the question of how a transformation in the industry is likely to come about. In the final chapter, we will put this in the wider context of society and the concept of motoring.

8 The Economic Implications of EOV Technology

8.1 THE ALTERNATIVE STRATEGIES AVAILABLE TO THE VEHICLE MANUFACTURERS

Our EOV concept is desirable from an environmental perspective, and several trends in the industry and the market appear to be pointing in its direction. Therefore, it is perhaps prudent to consider the implications of this. The EOV as set out in Chapter 7 argues for a major shift in the basic car concept and in the technology needed to build it. In addition, it implies lower production volumes in order to move towards a more sustainable motor industry. The major change, however, involves the abandoning of the Buddist paradigm (Chapter 4). The vehicle manufacturers have three basic alternative strategies open to them:

- continued enhancement of existing technologies
- exit from the industry
- transition to EOV technologies, guided by an EOV regulatory framework

With respect to the enhancement of existing technologies, the scope for further improvement in both product performance and manufacturing technology should not be overlooked. We have established in previous chapters that one of the major problems of the Buddist paradigm involves the very high investments required to run the system. This was true from the start; Morris, for example, spent no less than £120 000 in the late 1920s for the design and dies in tooling up for his first all-steel saloon (Ware, 1976: 75). Since then, the situation has only become worse as car body structures have become more complex. Manufacturers now have to comply with increasingly stringent crash test regulations, and with the requirements of cost reduction and more automated body build. Table 8.1 summarises the entry cost for traditional car manufacture.

The thrust of developments in steel automotive production technology has been to increase flexibility both in terms of product types and absolute volumes, to reduce cost and to drive down the break-even point at which the manufacturer is able to return a profit. At the same time, market pressures have continued to force a higher rate of new model introductions, which places great strain on the strategy of recovering investments over long production runs.

Table 8.1 Entry Cost for Budd-type Car Manufacture

Item	Cost (£ millions)
Press-shop	100
Press tooling	20–65
Body-in-white	50–100
Paint shop	200–250
Total	370–515

Source: CAIR, FT (19 March 1996).

These high costs are not only due to the specific hardware required to run the Budd system, but also to the product development, which is an expensive and time-consuming process. Even the lean Japanese producers have been unable to reduce the lead time for a new car below around three years. Compare this with a company such as TVR, which can develop a new body in a matter of months at minimal expense and operate profitably on total combined volumes of a mixture of models of less than 1000 a year (Figure 8.1).

The costs of the development of a steel monocoque can be recovered in a number of different ways. Clearly, building high annual volumes allows a

Figure 8.1 British Sportscar Maker TVR can Develop a New Body in a Matter of Months at Minimal Expense

Table 8.2 European Product Cycles: Volume Producers

Company	Model	Introduction (year)	Cycle (years)
Fiat	500	91, 99	8
	127/Uno/Punto	71, 83, 93	12, 10
	Ritmo/Tipo/Bravo	78, 88, 95	10, 7
Ford	Fiesta	76, 89, 95, 99	13, 6, 4
	Escort	68, 75, 80, 86, 90, 95	7, 5, 6, 4, 5
	Cortina/Sierra/Mondeo	70, 76, 82, 93, 99	6, 6, 11, 6
	Scorpio	77, 85, 94, 98	8, 9, 4
GM Opel	Corsa	82, 93, 97	11, 4
	Kadett/Astra	75, 84, 91, 96, 00	9, 7, 5, 4
	Ascona/Vectra	75, 88, 96	13, 8
	Rekord/Omega	77, 86, 94	9, 8
PSA	Citroën AX/Saxo	86, 96	10
	ZX	91, 96	5
	GS/BX/Xantia	70, 82, 93, 99	12, 11, 6
	DS/CX/XM	55, 74, 89, 97	19, 15, 8
	Peugeot 106	91, 99	8
	205	83, 97	14
	309/306	85, 93, 98	8, 5
	504/505/405/406	73, 79, 87, 95	6, 8, 8
	604/605	75–86, 89, 97	11, 8
Renault	Twingo	92	
	R5/Clio	72, 82, 90	10, 8
	R14/R19/Megane	76, 88, 95, 00	12, 7, 5
	R18/R21/Laguna	78, 86, 93	8, 6
	R20,30/R25/Safrane	75, 83, 91, 99	8, 8, 8
	Espace	84, 91, 96	7, 5
VAG	VW Polo	75, 81, 94	6, 13
	Golf	74, 84, 91, 97	10, 7, 6
	Passat	73, 88, 97	15, 11
	Scirroco/Corrado	74, 81, 88	7, 7
	Audi 80/A4	72, 78, 86, 94	6, 8, 8
	100/A6	76, 82, 90, 97	6, 8, 7
	V8/A8	88, 94, 00	6, 6
	Seat Ibiza	84, 93	9
	Cordoba	85, 94	9
	Toledo	91, 99	8
Average			8.13

manufacturer to achieve the required economies of scale relatively quickly. Most US and Japanese producers and several European firms have managed this reasonably well. However, if this is not possible — and we argue in Chapter 2 that market pressures are militating against high volumes — making smaller volumes over a longer number of years can also allow a manufacturer to achieve lifetime economies of scale. This is the reason why smaller specialist producers

Table 8.3   European Product Cycles: Specialist Producers

Company	Model	Introduction (year)	Cycle (years)
BMW-Rover	3	66, 75, 90, 98	9, 15, 8
	5	74, 88, 95	14, 7
	7	77, 86, 94	9, 8
	Rover Mini	59, 00	41
	Metro/100	80, 90, 99	10, 9
	200/400	89, 95/6, 00	6, 5
	Montego/600	84, 93, 98	9, 5
	SD1/800	76, 86, 98	10, 12
Jaguar (Ford)	Jaguar XJ6	68, 86, 94	18, 8
	XJS/XK8	75, 96	21
Mercedes-Benz	190/C	82, 93, 99	11, 6
	W123/124/E	76, 84, 94	8, 10
	S	72, (86), 91, 98	14, 6, 7
	SL	89, 98	9
Nedcar (Volvo)	300/400/S, V40	76, 88, 96	12, 8
Porsche	Boxter	96	
	924/944/968	76, 91	15
	911	63, 97	34
	928	77–95	18
Saab (GM)	Saab 900	78, 93, 00	15, 7
	Saab 9000	84, 97	13
Volvo	140/200/850	67, 74, 91, 99	7, 17, 8
	700	82, 98	16
	900	90, 98	8
Average			11.8

such as Jaguar or Volvo tend to keep their models in production for much longer than the volume producers do with their most popular models. Tables 8.2 and 8.3 show this by giving the years when certain models were replaced.

Europe differs from North America in this respect. In the US, the market is dominated by only three local manufacturers, which are therefore able to sell vehicles in much higher volumes. This makes it much easier for GM, Ford and Chrysler to achieve minimum economies of scale on each or at least on most of their models. In Japan, too, although there are more manufacturers, their worldwide expansion over the past few decades has allowed them to sell large volumes of their most popular models worldwide. In order to catch up with western producers in terms of product, the Japanese producers adopted a four-year cycle for their most popular models.

Although models such as the VW Golf and the Fiat Punto do reach similar high annual volumes, in Europe volumes per model are generally much lower. This is especially true for the specialist producers such as BMW, Mercedes, Volvo, Saab or Jaguar. In addition to running longer model cycles, these producers have been unique within the context of the world motor industry in being able to practise cost recovery (Williams et al., 1994). This means that their volume disadvantage has been offset by the fact that customers have been willing to pay a price premium for their products in view of their rarity and prestige image. The manufacturers of these lower-volume specialist cars have therefore been able to pass on the extra costs resulting from their lower volumes to their customers.

The generally longer product cycles of the specialist producers have in a sense become part of the quality image in that a long product cycle leads to lower depreciation (a 6-year-old Volvo or Mercedes can look almost the same as a new one), as well as a general feeling on the part of the market that the car must have been right to start with and therefore does not require early replacement. Long product cycles may also slow down the process of product innovation in many areas, despite the fact that changes in powertrain and specification can be made within an existing bodyshell.

The issue of product cycles came to the fore in the 1980s when western producers increasingly felt the need to match the four-year cycles of the Japanese. Since then, western car makers have moved gradually towards shorter cycles, as illustrated in Tables 8.2 and 8.3. Automotive forecaster DRI has calculated that European model cycles are declining from just under 10 years in 1985 to 6.5 by 1999 (Sewells, 1995). On the other hand, the Japanese are now extending their cycles in order to reduce costs and because the main objective behind the four-year cycle (i.e. catching up with the West in terms of product technology) has now been achieved. However, there is another reason for the slowdown in model replacement in Japan, as Heller (1996: 5) explains:

> The quicker the model changes, they now believe, the less time is available for originality and creativity. The drift of cars, no matter who makes them or in which country, into becoming look-alikes and me-toos is one consequence. Breakthroughs like mini-vans [people-carriers] and Miatas [Mazda MX-5] were look-differents and me-differents. The pressures of a forced development pace will not generate such individuality.

In practice, Tables 8.2 and 8.3 understate the length of time for which products remain in production. This is because of the high degree of 'carry-over' which may occur, i.e. the use of components from one model generation in the next model generation. There is also difficulty in establishing the difference between an external 'face-lift', in which a few large exterior body panels are changed, and a 'new' car. Carry-over is a direct consequence of the investment cost of the all-steel body. New model introductions between 1994 and 1999 are set out in Table 8.4.

It is clear that Budd technology provides little flexibility in terms of volumes and that its economies of scale force the industry into high volumes. The advantage of moving to plastics or composite technology, on the other hand, is that this is not one homogeneous technology. There are different types of plastic

Table 8.4 New Model Introductions 1994–99 by Segment, Europe

Segment	Typical model	No. of new model introductions
A	(Ford Ka)	6
B	(Ford Fiesta)	9
C	(Ford Escort)	16
D	(Ford Mondeo)	25
E	(Ford Scorpio)	30

Source: Sewells (1995).

and composite body technology, each with different minimum economies of scale. This is one of the main attractions of a move to plastic or composite technology. As the market increasingly demands visible differentiation in cars, the model ranges of manufacturers can become rather fragmented, with each variant made in fairly small volumes. Segmentation is constantly changing, and is increasingly developing niches and sub-segments. This means that any technology that allows a manufacturer to escape the tyranny of Buddist economies of scale is increasingly attractive, especially in Europe where an increasing variety of models are chasing a limited number of buyers.

The advantages of steel

The use of shared platforms (see Chapter 2) and components are both part of a strategy of 'intensification', whereby the vehicle manufacturers are seeking to reduce cost and/or increase flexibility in the face of market saturation and fragmentation. Huge efforts are being made to adapt Buddist technology to contemporary demands. We have already drawn attention to the ULSAB programme whereby leading steel companies have collaborated to improve the performance of the all-steel body (see Chapter 4). In the realm of steel processing technology, that used in forming and fabricating all-steel bodies, the introduction of integrated transfer presses has increased both the speed of parts production and reduced the time required to change from one part to another (Nelson & Metzger, 1995; Wells & Rawlinson, 1994b). At present, the industry is searching for die technology that would allow an intermediate investment between the cheap and low-cost prototype dies, and the very high-cost 'hard steel' dies used in volume manufacture (Rawlinson & Wells, 1996). However, one of the most significant advantages steel has as a material is the depth of knowledge available about how it performs, especially in the automotive industry. This has helped the continuous improvement and development of steel itself (Anon., 1992; Chatterjee, 1995), but also of the technologies which underpin steel processing in the automotive industry (Figure 8.2).

Thus, it is not surprising that considerable efforts have been made to reduce fixed costs and increase flexibility in steel body construction processes. In terms

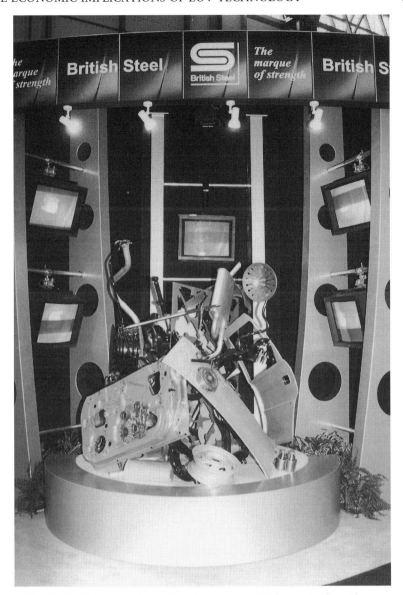

Figure 8.2 Steel Makers such as British Steel are Fighting Back with New Steel Technologies

of forming processes, there has been a drive to reduce the number of panels per vehicle and reduce the number of dies per panel; in effect, reducing the total number of press-blows required per vehicle. The use of modern integrated transfer presses has increased the rate of press-blows delivered, while simultaneously increasing the speed at which dies can be changed. The largest such press in Europe has recently been installed at its Sindelfingen plant by Mercedes and is capable of pressing complete monosides (i.e. a complete car body side panel from a single blank) at the rate of 14 parts per minute. Moreover, the use of tailored blanks — whereby a blank is made up of several smaller sheets welded together — offers further opportunities for parts consolidation, higher material yield and material optimisation.

In terms of joining, the use of spot and seam welding robots is now almost universal for steel body fabrication. Robots offer considerable flexibility in terms of the operations they may perform, and thus can be used for other models or operations other than welding. For smaller parts and subassemblies, fixed multiwelders are generally used — though innovations in Japan have led to the development of multiwelders with interchangeable heads. While the use of coated steels has created some extra difficulties and costs in fabrication, in general joining costs per vehicle have reduced, while quality and overall body performance (dimensional accuracy, stiffness, etc.) has increased. One key bottleneck here is the framing of the body for joining purposes. Traditionally, this has been done on fixed, model-specific jigs which were expensive to manufacture and which had to be discarded for subsequent models. However, in recent years a new approach to flexible body framing has been developed by companies such as Nissan and Volvo.

The Nissan IBAS (Intelligent Body Assembly System) line is 65 m long and capable of producing one vehicle every 45 s, i.e. around 220 000 vehicles per annum (Abe et al., 1995). Initially, seven major subassemblies, including the floorpan, roof and body sides, are pre-assembled in order, without welding. The body is then passed to a so-called NC locator which acts as a programmable parts-holding jig. The NC locator has 51 robots integrated into one machine (35 for positioning parts, 16 for welding). While parts are held by multi-position robots inside the NC locator, they are tack welded in 62 locations. The body is transferred to a body measurement station, which measures 26 locations for the standard in-line quality test. However, the measurement station may also undertake more detailed measurements, as required, when the line is stopped.

In their evaluation of the system, Nissan considered that average investment was 16% higher than for standard body framing fixtures, but that a cost saving of 70% can be realised at model change-over time. Nissan expect to generate savings of ¥20 billion over the next five years. Body accuracy was higher than with the conventional approach, while a respectable capacity utilisation figure (excluding planned downtime) of 97% was obtained.

In terms of working practices, intensification is occurring by changing shift patterns, leading to longer working shifts and reduced planned 'downtime' when the assembly plant is not producing. In so doing, the vehicle manufacturers effectively reduce the payback period on investments. Moreover, volume flexibility can be achieved by adding (or removing) shifts. It is the workers who must bear the brunt of these attempts to sustain the Buddist paradigm (Rawlinson & Wells, 1996). It might be added that the vehicle manufacturers and their suppliers are adept at securing state support for new investments, in both existing locations and new ones, and this further insulates the industry from the true cost of the existing paradigm.

Finally, with respect to sustaining the existing paradigm, it is important to consider the relationship between capacity, supply and demand. There are two distinct strategies here. The volume vehicle manufacturers will seek to install capacity to meet the peak of demand, and seek to maximise profits through higher margins once the break-even point has been reached. Profits are made during the peak of the demand cycle but, as was shown in Chapter 1, large losses are made at the bottom of the demand cycle. The volume vehicle manufacturers thus seek to make enough profits in the good years to sustain them in the bad — in the hope and expectation that cash reserves will be sufficient until the next upswing. As we suggested in Chapter 1, this rather depends upon a resurgence in demand occurring. So, for the volume vehicle manufacturer using Buddist technology, every downswing in the demand cycle is a gamble.

On the other hand, the small specialist producers install production capacity well below peak demand, and ideally at the level of demand which pertains at the bottom of the demand cycle. Then, in times of high demand, the specialist producers merely increase the waiting time for customers, and thereby actually enhance their 'prestige' image. An extreme example of this approach is the MG 'F' sports car. Its body is made of steel, but is assembled in a low-automation, highly labour-intensive and flexible system by Mayflower for the Rover Group. The system has a capacity of only 20 000–25 000 a year. Final assembly is carried out by Rover using stock components. Since its introduction, the car has been in short supply, thus optimising the investment in normally expensive steel body technology which in this case constantly runs at or near capacity. This car uses installed investments for Rover and Mayflower as a whole, and is thus able to stretch the volume limits of the Budd paradigm as far down as possible.

As noted above, a further alternative for the vehicle manufacturers is to exit from the sector entirely. The question of exit costs is often neglected (Raff, 1994) but, in the case of the automotive industry, these are very high. There are two main problem areas: staff and capital equipment. With respect to staff, mass redundancy is an extremely expensive option; one of the reasons that GM recorded such large losses in the early 1990s was that it closed a large number of plants at that time (Wells & Rawlinson, 1994a). It is notable that in France, PSA and Renault announced in 1996 that they would seek government support for a redundancy programme intended to lower the average age of the workforce. Under the plan, some 40 000 jobs would be lost at an estimated cost of US$6 billion by the year 2001. At the same time, 14 000 young workers would be hired, leaving a net reduction of 26 000 jobs (Farhi, 1996).

With respect to capital equipment, it is the case that some of the machinery used in automotive production (presses and robots, for example) could find applications elsewhere. Dedicated equipment such as the paint shop, and the jigs and fixtures used in the assembly process, together with any dies, would not have a ready market. Even in the case of presses and welding equipment, it is the case that the automotive sector is one of the primary users of such equipment and, again, these assets would command a negligible value. There is already substantial over-capacity in presses, for instance, with large numbers of second-hand tandem presses being made available as a result of the wave of investment in integrated transfer presses. Most of this equipment has been exported to emerging automotive production locations in, for example, Korea and Brazil. So it is reasonable to suggest that the immediate exit costs for a vehicle manufacturer are extremely high. Of course, the wider social costs could be much higher.

8.2   THE THREAT FROM ALUMINIUM

Aluminium has a long history of use in the automotive sector: in the 1920s and 1930s it was common for luxury vehicles to have aluminium bodies. Despite this, and despite further use in, for example, racing machines and, briefly, in mass production by Panhard in the 1940s, the widespread adoption of aluminium did not occur. The aluminium industry has been actively seeking to promote the further use of its material in vehicles and, since the 1980s, has made a clear attempt to win further use of aluminium in vehicle bodies (Kewley et al., 1987; Nardini & Seeds, 1989; McGregor et al., 1992; AEA, 1995; Aluminum Association, 1995; Williams, 1996). Table 8.5 shows a material cost per body comparison.

In the material cost matrix shown in Table 8.5, an average of 75% material yield is assumed, except in the case of the spaceframe-type design where yield from extrusions and castings may be expected to be higher (in the order of 90%), given that material may be simply recycled on a closed loop within the production process. In all cases the body includes hang-on panels (i.e. doors, bonnet, bootlid, fenders). The aluminium 2 body assumes 110 kg of sheet, 66 kg of extrusions and 44 kg of castings. While some of the casting alloys may be more expensive per kilogram than that for sheet, generally extrusion alloy prices will

Table 8.5   Materials Cost Matrix

Material	£/kg	kg/car	Material yield (%)	kg/car total	Total cost per car (£)
Steel 1	0.45	315	75	420	189.0
Steel 2	0.50	315	75	420	210.0
Aluminium 1	2.00	252	75	336	672.0
Aluminium 2	2.00	220	75–90	250	500.0

Note: steel 1 = ordinary CR steels; steel 2 = fully galvanised or other premium steels; aluminium 1 = monocoque or unibody design; aluminium 2 = Audi A8 type design. Does not include value of scrap in production or post-use.
Source: CAIR.

be cheaper. The aluminium 1 body assumes just 6000 series external panel quality prices, but in practice a substantial proportion of the body (perhaps 40%) could be built using 5000 series aluminium, which would be cheaper.

Steel, even with a fully galvanised body, is certainly cheaper than aluminium on a per vehicle basis. We would expect that the relative price advantage of steel could be further increased by continued improvements in material yield arising from tailor-welded blanks, better die design, parts consolidation, better quality steels and reduced line scrap. All of these measures would improve the material yield. In contrast, a material yield of 75% for aluminium sheet is still generous; we would expect the scrap rate for sheet aluminium to be higher than that for sheet steel. Additionally, the prospect of material yield improving in the future is far less for sheet aluminium than for sheet steel.

Both steel and aluminium show long-term declines in real price per kilogram, and a slight closing of the price differential between the two over time. It is reasonable to expect both trends to continue into the future, i.e. a slight convergence in the context of long-term declines in real prices. In net terms this means that, in material cost, aluminium will demand extra expenditure of at least £1/kg weight saving achieved. In the example used in Table 8.5, using aluminium as a substitute for steel (in the spaceframe design) saves about 170 kg at a cost of £311 per vehicle body (i.e. about £1.8/kg weight saved).

The traditional preference for steel arose out of its relatively abundant supply, the strength and formability of steel, and the development of a reasonably robust means of joining panels together by spot welding. Subsequently, steel has provided a benchmark for other material characteristics, such as longevity in use, surface finish, repairability, etc. There can be little doubt that aluminium does not compare favourably with steel in terms of most criteria or characteristics desired. Formability and yield strength are both substantially below most steels, for example. It is only by virtue of its relatively lower density that aluminium has become attractive as a material for building vehicle bodies. The main advantages and disadvantages for aluminium, steel and plastic are summarised in Table 8.6.

There has been a growing use of aluminium for external body panels as a means of weight reduction, particularly in the US market — though there are also European-produced aluminium cars (e.g. Land Rover Defender and Discovery, Audi A8 and the Lotus Elise chassis) as well as those using occasional external panels (e.g. BMW M3 and Volvo 960). Japan has only one all-aluminium car — the Honda NSX — though several other models which are exported to the US market have some aluminium panels. The position for cars produced in the US is shown in Table 8.7. The table shows that a total of 840 000 hoods/bonnets and 600 000 deck/bootlids will be in production in 1997.

As a rough guide, the industry considers that every 1% reduction in weight can improve fuel consumption by 0.6%. While we have concentrated on aluminium use in vehicle bodies, it is already the case that aluminium has displaced cast iron or sheet steel in a range of applications, including suspension systems, engines and cylinder heads, wheels and brake components. The partial breakthrough for

Table 8.6   Advantages and Disadvantages in Each Major Material Group Used for Vehicle Bodies

Material	Advantages	Disadvantages
Steel sheet	Formability, joinability, paintability, production tolerant, cost	Weight, corrosion, easy to damage
Sheet aluminium	Weight, energy absorption	Formability, joinability, very easy to damage, production intolerant
Thermoset	Formability, paintability, repairability, weight	Production intolerant, cycle times, colour stability, surface finish
Thermoplastic	Formability, flexibility, robust in use	Cost, cycle times, thermal expansion

Source: CAIR.

Table 8.7   US Cars Currently in Production or Soon to Start Production that Use Aluminium Panels

Model	Part	Annual volume
Lincoln Town Car	Hood	100 000
Mercury Grand Marquis	Hood and deck lid	60 000
Ford Crown Victoria	Hood and deck lid	140 000
		(200 000)
1996 Ford Taurus	Deck lid	400 000
1996 Mercury Sable	Deck lid	(total)
Oldsmobile Aurora	Hood	40 000
Buick Riviera	Hood	30 000
1996 Ford full-sized pick-up	Hood	350 000
1996 GM APV minivans	Hood	80 000–90 000
1997 Buick Park Avenue	Hood	30 000
Total Hoods 1997		840 000
Total Deck lids 1997		600 000

Source: CAIR.

the Alcan AIV (aluminium-intensive vehicle) system was the collaboration with Ford in the US to produce the Synthesis 2010 car. This experimental car was made as a weld-bonded aluminium monocoque with aluminium closure panels. Ford and Alcan claim that the car weighs 900 lb (408 kg) less than if it were made out of steel. Following on from their Synthesis experiment which was aimed at investigating technologies for the longer term, Ford decided to build a fleet of 40 test cars. The aluminium-intensive vehicles (AIVs) weigh 400 lb (180 kg) less than the equivalent steel-bodied Taurus/Sable cars on which they are based (Table 8.8). Ford and Alcan claim that further secondary weight savings of the order of

Table 8.8 Weight Savings Possible in the Ford/Alcan AIV Project

Weight savings in the Ford AIV	Steel	Aluminium	Weight saved (lb)	% saved
Body structure	596	320	276	46
Hood, deck and fenders	90	38	52	58
Front and rear doors	132	79	53	40
Total body and closures	818	437	381	47
Total vehicle	3275	2894	381	12

Source: Ford & Alcan (1994).

300 lb (135 kg) could be achieved through downsizing the Taurus/Sable powertrain, transmission, braking, steering and suspension systems.

Additional secondary savings arise because of the lighter body shell. For example, the suspension can be made lighter and the engine can be made smaller, yet still offer the same performance as a heavier car. The ability to generate secondary savings is a significant advantage that aluminium has over the use of a conventional steel body. As illustrated, the lightweighting of the body and closures alone results in a total saving of only 12%. The secondary weight savings in the Ford AIV project generated another 307 lb (140 kg) of weight loss, to give a total weight reduction of 688 lb (312 kg). The breakdown of the secondary weight reductions are shown in Table 8.9.

Supply of aluminium materials

Once again, the industrial infrastructure already in place for steel forms the benchmark against which the aluminium and plastic supply chains must

Table 8.9 Secondary Weight Savings on the Ford/Alcan AIV

Primary	Weight saving (lb)	Secondary	Weight saving (lb)
Aluminium structure	276		
Aluminium closures	105		
		Powertrain	120
		Suspension	65
		Brakes	24
		Steering	11
		Fuel system	20
		Wheels and tyres	26
		Exhaust system	9
		Drivetrain	32
Total saving	381		307

Source: Ford & Alcan (1994).

compare. Thus, if the knowledge and experience base for processing sheet aluminium is weakly developed in vehicle assembly, it is even weaker in the independent presswork sector. Similar observations may be made with respect to, say, the machine tool industry: it is significant that Audi had to source their extrusion-bending equipment from the USA, for example.

The aluminium industry is usually divided into primary and secondary production. Primary production entails the conversion of bauxite ore into pure alumina, which is then smelted into pure aluminium, a process which demands high levels of energy input. Primary producers of aluminium often have upstream vertical integration interests in bauxite mining, shipping or power generation. Major bauxite deposits are relatively few in number, with large reserves in the former Soviet Union, Australia and Jamaica, for example. Secondary production uses scrap aluminium alloys to produce new ingots; the energy demands are much lower for secondary aluminium, but there are added costs in sorting and purifying scrap. The major primary producers generally have further interests in secondary production, but there are also a large number of smaller independent producers. In terms of downstream processing, aluminium is often used in extruded or cast form where a substantial independent industrial infrastructure exists, but also where some of the primary aluminium producers have vertically integrated downstream interests.

Historically, primary aluminium production was associated with the North America region, and especially the companies Alcan and ALCOA. North America in the 1970s accounted for almost half of world primary aluminium production; Alcan and ALCOA combined accounted for about one-third of production. Since then, important changes have included the following:

● Decreased control by Alcan and ALCOA over the total market in the face of new entrants.
● Geographical shifts in production, mainly towards locations offering cheap or subsidised electricity.
● A general expansion in capacity of the order of 3% per annum.
● Stronger growth in secondary aluminium production capacity, especially in industrialised countries.

The leading primary producers have remained the same: Alcan, ALCOA, Reynolds, Kaiser, Pechiney, Alusuisse and Norsk Hydro — but their overall share of the market has declined slightly over time. Total global primary aluminium capacity in 1995 amounted to about 20 million tonnes; that for secondary aluminium, about 7 million tonnes. The sudden availability of aluminium production capacity in the former Soviet Union has had a destabilising impact on markets in Western Europe since the late 1980s. In the search for lower-cost production locations, the aluminium industry has been forced to expand away from the major industrialised regions. There is effectively no primary aluminium production capacity in Japan today, while that in Western Europe depends upon either cheap hydroelectric power or the surpluses generated by nuclear power.

The overall picture on the supply side is, then, a very narrow supply chain in which, increasingly, capacity is remote from the major demand locations, where there are relatively few players and where only a proportion of capacity could be used to make automotive body sheet. These factors are important because they define the limits to which aluminium can capture further market share. To take the most extreme case, it is a simple matter to calculate the approximate required capacity for automotive body sheet aluminium if all European vehicles were built with aluminium rather than steel. At 336 kg per car, it would require just over four million tonnes of automotive body sheet capacity to produce 12 million cars in Europe — this is approximately equivalent to the entire primary aluminium production capacity in Europe. Given that automotive body sheet accounts for only a small part of aluminium (secondary) output, primary capacity and automotive body sheet capacity would both have to be expanded by at least four million tonnes. It takes three to five years and at least £100 million to install an aluminium automotive body sheet plant of a capacity of 100 000 t per annum. Thus, the total investment required of the aluminium industry would be in the region of £4000 million at current prices simply to expand automotive body sheet capacity to the required volume.

With current levels of aluminium usage compared with the highly 'lumpy' capital investments required to install new capacity, the problem for the aluminium industry is to expand that capacity in line with demand. On the one hand, vehicle assemblers need to feel assured that capacity will be available to produce the sheet they require. On the other hand, aluminium producers do not want to invest such large sums speculatively, in the hope that demand will arise, and thereby risk under-utilisation and a failure to recover costs. This applies equally to the downstream processing of aluminium. A key feature of the relationship between ALCOA and Audi, for example, was the investment ALCOA had to make in Soest, not just because they had the expertise to undertake the work, but also as a clear indication of a commitment to the A8 project. Again, this is a further indication of the aluminium industry taking a 'loss leader' view of gaining market share in the vehicle body — though if ALCOA does win more business for its Soest plant, then the long-term gains should be substantial.

Aluminium producers have a range of strategies with respect to increasing usage of their material in vehicle body structures for European production. ALCOA has, of course, been promoting the spaceframe concept, with a strong bias towards use of extrusions and castings rather than sheet aluminium. Alcan has, in contrast, been associated with aluminium monocoque concepts (following early work with what is now Rover in the UK), and has sought to move out of the extrusion business altogether — mainly on the basis that extrusions offer little by way of value-added activities. Norsk Hydro has no automotive body sheet capability and has focused attentions on extruded structures through its subsidiary in Denmark (Figure 8.3). This subsidiary has been involved in the promotion of aluminium for vehicle bodies, as in the case of the MOSAIC

Figure 8.3 Norsk Hydro Supplies the Aluminium Extruded Chassis for the Lotus Elise, Which has GRP Body Panels

project, for example. In contrast, Alusuisse is a major supplier of external finish quality automotive body sheet, but does not appear to have played such a proactive role as ALCOA, Alcan and Norsk Hydro in promoting aluminium use in vehicle bodies. The other major European producer, Pechiney of France, also has a direct involvement in the automotive sector through subsidiary companies: its main activities are in castings (mainly drivetrain) and, again, has not been a noticeable force in terms of promoting aluminium for vehicle bodies. ALCOA has been especially aggressive in its attempts to win a greater market share in vehicle body structures, and has been reported as offering US$1 billion to support the first volume application of aluminium. The overall strategy appears to be one of buying market share now, in the hope that profits can be recovered with greater volumes in the future.

To respond to the plastics industry, several metal producers have developed sandwich composite materials, which combine metal with plastics. The Dutch firm Hoogovens has developed Hylite™, a sandwich material of two very thin layers of aluminium with a thicker layer of polypropylene in between. The advantage is in weight and stiffness, although formability is not in the same league as aluminium alone. The price of Hylite™ is comparable with aluminium. The steel industry has also introduced a number of different sandwich materials which work on the same basis. Such laminate steels can be used for weight reduction and sound deadening in cars. It cannot be deep pressed due to formability problems. Welding also presents difficulties, while the laminate is not stable at the elevated temperatures found in the paint process. Both materials also present recycling problems, although these can be overcome. Hoogovens see the main application of Hylite^{TM} where flat surfaces are used, such as in floorpans. The roof and bonnet of the NedCar ACCESS, for example, are made of this material, while laminated steel is used in three panels of the ULSAB prototype vehicle.

Of course, none of the aluminium suppliers can claim experience in delivering high volumes of consistent-quality automotive body sheet to the vehicle manufacturers. With steel suppliers, there is a situation of mutual dependency with the vehicle manufacturers, i.e. while the automotive sector is dependent upon wide strip steel as the most important material input into the production process, it is equally the case that steel producers are dependent upon the vehicle manufacturers as high-volume consumers of value-added sheet steels. In the case of the aluminium industry, the automotive sector does not represent a core market, especially at the level of wide strip aluminium. Vehicle manufacturers prefer their suppliers to have a core commitment to the automotive sector; a shared destiny is seen as a means of reducing predatory pricing by suppliers.

8.3   THE PLASTIC ALTERNATIVE

One of the advantages of the existing way of making cars is that there is a widespread, well-developed knowledge base for the technologies concerned, with a massive infrastructure of materials and components supplier firms, equipment suppliers and service-sector operations. Compared with this infrastructure, that for alternative materials and power sources appears less well developed. There is some merit in this view, but it is also the case that there is considerable experience with using alternative materials and technologies arising out of a long history of use.

Composite bodies date back to the early days of the motor car when a steel chassis often supported a body made of an ash timber frame clad with steel or aluminium panels. Plastic composites came later. Reinforced-plastic technology was developed for a range of products in the 1930s. During World War II it was used for structural applications. Automotive applications emerged soon after the war, and the first examples appear on low-volume US kit cars from 1946 onwards. In these cases the material is 'hand uplay' glassfibre-reinforced plastic (GRP), which still dominates the automotive plastic body on low-volume applications, such as Marcos or TVR in the UK. The first European application was on the Berkeley Caravan, built from 1948; the company later switched to GRP bodied cars. Higher-volume applications appeared in the US in the early to mid-1950s with the Kaiser Darrin of 1952 (of which a few hundred were built) and the Chevrolet Corvette of 1953, which was significantly more successful. Table 8.10 summarises developments in plastic body technology.

The first European car to be designed for higher-volume production was the Volvo P1900, of which only 67 were actually built between 1955 and 1957. By

Table 8.10 Advent of the GRP Body, 1946–55

Country	1946	1947	1948	1949	1950	1951	1952	1953	1954	1955
USA	Comet, Bobbikar		Airway	Imp			Kaiser Darrin	Chevrolet Corvette	Darrin Aurora, Scorpion	American Buckboard
UK			Berkeley caravan					Allard Clipper, Jensen 541	Ashley, RGS, Fairthorpe Atom, Kieft1100	Elva, Turner
West and East Germany									Bruetsch	Spatz, Trabant
France								SOVAM, Marathon, Reac	DB	Alpine 106, Avolette, Meyra
Norway, Denmark and Sweden								DKplasticbilen	Volvo P1900	Troll
Switzerland									Agea-de-Toledo	Belcar
Czechoslovakia										Mobil

Source: CAIR.

1954/5, GRP technology had become widely accepted as most suitable for low-volume specialist applications, and quality gradually improved over the next decade. Japan's first GRP car did not appear until 1957 in the shape of the Fuji Cabin, a microcar built by Fuji Heavy Industries, now makers of Subaru cars.

The next major advance came with the GRP monocoque. This was pioneered on the Lotus Elite of 1959. This only carried reinforcements, bonded in for the suspension pick-up points. However, this technology has had few followers. One of the few plastic monocoques currently available is that of the Welsh Darrian, a specialised rally kit car which uses a GRP monocoque with carbon-fibre reinforcements. There are a number of other applications, such as the experimental electric vehicles by Horlacher of Switzerland. The Neoplan Metroliner MIC is a city bus with a GRP and carbon-fibre monocoque body, weighing a mere 989 kg for the 8-m N8008 version. This bus is currently in production and selling in some numbers, especially in Germany.

8.4 RTM VERSUS SMC

Modern thermoset materials used in automotive body applications can be divided up into two main categories: resin transfer moulding (RTM) and sheet moulding compound or composite (SMC). The vehicles outlined in the preliminary historical section all had bodies made using a technique that can be described as GRP hand lay up. More sophisticated techniques, such as the VARI (vacuum-assisted resin injection) system, developed and used by Lotus, involve pumping the resin in (RTM) or sucking it in by creating a vacuum (VARI). The latter technique allows a more even and consistent filling, and hence higher quality.

Plastics technology allows the consolidation of panels, i.e. by combining panels that in sheet metal would have to be separated in order to allow for the limitations of the pressing process. As a result, the total number of body panels can be reduced; in some cases quite dramatically. For its two-model range in the 1980s, the Esprit and the Excel, Lotus used essentially two moulds for each body. The bottom half of the body and the top half each came out of its own single mould. This technique allows great dimensional integrity and accuracy. However, the investment in full body-size moulds is higher than for smaller panel-sized ones, while the risk involved in having four moulds for your entire model range are considerable. Serious damage to one mould could affect production dramatically. Still, this approach suggests that entry level costs are an order of magnitude lower than those for mainstream steel body technology — an interesting feature because it suggests the possibility of new entrants into the automotive industry. Thus, plastics production technology allows firms to gain a foothold in the market with relatively limited investment, albeit at low volumes.

Even in the fastest RTM processes, parts require at least 4 min of curing time within the mould. In addition, time is taken for the full hardening process outside the mould. Considerable research effort is now being directed at improving the manufacturing performance of RTM, for example in Ford (Bonnett et al., 1995).

Within the curing time, a large number of sheet metal parts could have been pressed. The only answer would be to increase the number of moulds making the same part; however, this increases the investment considerably, while taking up much more room and complicating the whole production process, probably adding considerably to the labour content of the whole process. This apparent cost may also be an advantage, in that it illustrates the possibility of incremental investments such that production capacity may be more closely aligned to actual demand.

One of the main hurdles for the acceptance of GRP as a mainstream body material has been this curing time. The plastic will take many hours to set to its required hardness. During this time the tools are often tied up, or a large buffer stock is created. Normal mass-production processes cannot accommodate this waiting period. In fact, all plastic producers are aiming to achieve as close as possible to a 1-min cycle time for their materials in order to fit them into the existing production processes of the car producers.

This disadvantage has to a large extent been overcome by the introduction of sheet moulding compound (SMC). SMC involves the production of a composite-reinforced resin sheet which is cured in that form and subsequently rolled. The next process is more like working from a coil of sheet metal in that the SMC is uncoiled under moderate heat, then blanks are cut and moulded in a press not unlike sheet steel or aluminium.

As the main curing time is separated from the production process, cycle times compare much more favourably with sheet metals. The curing time in press is 1 min 20 s at Matra, for example. This allows production per die of around 25–30 parts per hour. As a result, SMC has already been able to replace steel in several applications. ERF makes the entire outer cab for its trucks from SMC, while Iveco has gone a similar route for its latest Euro range. Iveco has one of Europe's largest SMC facilities at its Brescia plant in northern Italy. Another major user of SMC is Matra Automobile in France, who moved to SMC from GRP for the Espace people-carrier, which it designed and developed and now builds for Renault.

The initial investment is higher for SMC. Unlike RTM, SMC requires presses which are not unlike those used for sheet metal body panels, although generally of a lower power. In addition, heated dies (rather than moulds) of a much higher quality are needed. Unlike RTM moulds, SMC dies need to be strong enough to withstand the power of a press. However, SMC dies do not need to be of the same strength as a sheet metal die, although the chrome coating needed to get a good surface finish does add to the cost. SMC can therefore only be justified for larger volumes. Nevertheless, the dies used for SMC are around four to five times cheaper than those used for steel. In this respect the investment levels can be considered as lying somewhere between RTM and steel.

Compared with RTM, a move to SMC does have the advantage of considerable labour savings as the same labour-intensive surface treatments are not required. On the other hand, SMC is generally considered unsuitable for panels of more than $1\,m^2$. Beyond this, the integrity of the surface is more likely to become compromised, i.e. it is difficult to produce a smooth surface.

Thermoset materials do not appear to have any customer acceptance problems. Few owners seem to have been aware of the fact that the Citroën BX had several plastic skin panels and, if they were, they did not object. The bonnet, scuttle ventilator intake, rear hatch and C pillar panels were all made of SMC, while the bumpers were made of polypropylene. Many Espace buyers are also unaware of the unique construction of this vehicle compared with its competitors; many French buyers have opted for the Chrysler Voyager purely on the basis of price, whereby the Espace's price premium gives a certain 'cachet'. Because of this experience, even Mercedes-Benz is now experimenting with some plastic skin panels on production cars (see below).

SMC in the US

In the US, thermoset panels have grown tremendously in popularity over the past few years, as fears of tighter CAFE standards — and the threat of a spread of these to light trucks — has prompted greater interest in weight saving. Much of the resulting demand for lighter materials has been filled by SMC, and new quantum leaps in terms of viable scale of SMC production have been made. The world's longest-standing plastic production car, the Chevrolet Corvette, changed over from RTM to SMC in 1971 and, since then, SMC has become America's favourite material for weight-reducing body panels.

Some success stories for the SMC trade association, the SMC Automotive Alliance, which promotes the material among the Big 3 automakers, are the top of the range 1995 Lincoln Continental and Ford Windstar. The Lincoln Continental front wings are the first application of flexible SMC, giving some of the advantages of thermoplastic materials (see below). Other panels made of SMC on these vehicles are shown in Table 8.11.

Research in the US suggests a cost saving over aluminium of 50–55% per pound of weight reduction at volumes of 150 000 a year (Ward's Autoworld, 1995: 3). The SMC bonnet on the Continental is 25% lighter than its steel counterpart, while the Windstar takes SMC to a new high in terms of volumes, at

Table 8.11   SMC Panels on the Lincoln Continental and Ford Windstar

Lincoln Continental	Ford Windstar
Bonnet	Bonnet
Bootlid	Scuttle vent
Quarter panel extensions	Window frames
Sunvisor	Side skirt
Grille reinforcement (structural)	

Source: SMC Automotive Alliance (1995).

280 000 a year. Other mainstream models incorporating significant amounts of SMC are the Ford Mustang, the Ford Ranger, the Chevrolet Camaro, the Pontiac Firebird and the Chrysler Voyager.

As a result of this success, SMC use in automotive applications in North America increased by more than 25% from 1993 to 1995. The most popular panel to be replaced with an SMC item is the bonnet. In all, 22 models built in the US in 1995 used SMC bonnets, representing more than a quarter by weight of total SMC use in the vehicle industry, which amounted to 90 million kg in 1995. North America is also ahead in SMC recycling, with a special process developed by Phoenix Fibreglass of Oakville Ontario looking particularly promising. SMC is by far the most popular plastics technology for automotive panels in the US, taking 70% of the plastic panel market in 1994 – with suppliers such as GenCorp taking an active role in promoting further use (Kleese et al., 1995).

8.5   THERMOPLASTICS

Thermoplastics are already widely used in cars. Several cars have fuel tanks made out of them, while many of the fluid containers in cars are also made out of these materials. Other common applications include bumpers, spoilers and interior mouldings. However, their application in exterior body panels is still rare. The materials in question are known as engineering thermoplastics, or ETP.

One of the main proponents of this type of material for car bodies has been General Electric Plastics (GEP). Their European centre, based at Bergen op Zoom in the Netherlands, has made strenuous efforts to convince European manufacturers about the merits of these materials. GEP even operates a small vehicle design centre, which has shown concept cars, and regularly communicates with other design houses such as Pininfarina. The thinking behind this approach is that the materials choice should precede the design stage in that different materials require a different design approach. In terms of styling, thermoplastics allow a particularly flexible approach to shapes; these materials are far more forgiving than either steel or aluminium.

However, the only current applications of GEP's materials are as a replacement for steel panels. By 1996, three European cars were fitted with front wings by GEP: the Renault Clio 16V and Williams, the Renault Mégane Scenic and the Mercedes E500. These are prestige or performance models within their respective ranges, and weight saving was a key criterion. On the Clio, a weight saving of around 1.5–2 kg was achieved, by using GEP's materials for both front wings, instead of opting for aluminium. Truck producers have been more adventurous in their willingness to accept plastics in cabs; the Volvo FH, for example, contains around 65 kg of thermoplastics in the cab front.

Dupont supplied the one-piece plastic roof for the GM Impact. This is made of a thermoplastic known as XTC^{TM}, and it is claimed that this allowed a weight reduction of 15%. In addition, thermoplastics can easily be recycled, and Dupont has calculated that roofs can be made with between 25% and 100% recycled material. Bringing the roof up to a class A quality level is much easier than with SMC (Mom & van der Vinne, 1995: 47).

These panels have other advantages apart from weight saving and ease of recycling. Thermoplastics allow a high-quality surface finish to be achieved at minimal cost. In addition, this finish can largely be retained after an impact. Unlike metals, which lose their shape permanently even after a minor impact, thermoplastics will bounce back and return to their original shape. As a result, they are particularly attractive as a replacement for vertical skin panels such as wings and door skins. These properties are likely to attract the attention of the car insurance industry which is facing mounting costs from claims to repair steel panels. Aluminium offers few advantages here, as it is at least as susceptible to small dents as steel.

Another advantage is that as it is not bent in processing, it can be made with essentially a zero panel gap between adjoining panels, should that be required. All that is needed is a flexible mounting to allow the panels to move, and expand and contract, as required. In addition, GEP has, in co-operation with Renault, been able to reduce the cycle time for injection moulding to a competitive 50 s, while its Noryl GTX^{TM} resin is available in a version that allows on-line painting in paint ovens up to 170 °C.

However, thermoplastics are on the whole less suitable for structural applications. GEP is targeting its materials particularly for side panels such as wings, door skins, and the front and rear sections. In fact, they promote an approach whereby the distinction between the front and rear panels (or bumpers) and the wings disappears. Because of the nature of the material, a separate bumper is not necessary: the whole front outer skin can be used to absorb small bumps. Even the lights can be incorporated by using GEP's high-quality thermoplastic lens materials. In fact, the whole body could be made from a clear material, with the lights and windows masked off for painting. Such an approach would not only allow the stylist greater freedom, it would also greatly simplify the body structure, thus reducing cost.

The greatest problem with thermoplastics is the rate of contraction and expansion when they are exposed to heat or cold. This is far less of a problem with thermoset materials. Some SMC materials have an expansion and contraction rate not unlike steel. However, current thermoplastics used in automotive applications can expand as much as 10 mm per panel over a 120 °C temperature range. In addition, engineering thermoplastics are relatively expensive. This is due to the fact that, beside a higher investment in equipment, most producers control the exact composition and process, and can thus command a price premium while retaining exclusive rights over their product. Unlike steel, aluminium or even SMC, these materials are not yet simple commodities.

RIM/RRIM/SRIM

Reaction injection mouldable polymer systems (RIM) in a sense combine some of the properties of thermoset and thermoplastic materials in that they have the structural advantages of the former with the mouldability and short cycle times of the latter. In essence, an isocyanate — with glass flakes and milled glass mixed in for reinforcement — is mixed with a hardener, and is injected into a mould to promote a rapid reaction and thus setting time. Catalysts are used to control the speed of the reaction and these are the key to the cycle time, which has been reduced to close to the crucial one minute.

RIM materials offer considerable scope in vehicle body applications, as some of them (SRIM) are suitable for structural applications. SRIM was, in fact, used in Renault's MOSAIC programme in conjunction with SMC. The Pontiac Fiero, built by GM in the 1980s, used a steel chassis with RIM outer panels. RIM offers the potential for greater temperature resistance. This makes it more adaptable than most plastics to on-line painting in paint shops designed to process steel bodies. Dow Chemicals has developed materials that will survive temperatures of up to 200 °C without significant damage.

New variants are adapted to automotive applications all the time. The Alfa Romeo Spider and GTV, for example, have a large front bonnet made out of a completely new material called KMC, a composite material consisting of polyester resin reinforced with 25% carbon fibre. The material is injection-moulded like 'conventional' RIM materials. The moulds, heated to 150 °C, are chrome-faced and pressurised after the material is injected. The hardening process takes around 40 s, allowing a cycle time well within the critical one minute. Alfa use a conductive primer on their KMC parts to allow painting with the rest of the steel body.

8.6   ECONOMICS OF PLASTICS

As manufacturers move towards greater product differentiation and more niche variants even on mainstream models, body variants begin to proliferate, leading to a larger number of relatively lower-volume bodies. This makes the high production runs needed to run a steel BIW process profitably increasingly difficult to achieve. Plastics may offer part of the answer, either in combination with steel or with aluminium. However, how do plastics/composites compete in a steel world?

Raw material costs typically account for 65–75% of the direct manufacturing cost of a typical body panel. The remainder is taken up with labour, capital investment, energy, tooling, etc. The MOSAIC programme concluded that in making a Clio-type car in Clio volumes of around 1800 cars a day (around 400 000 a year), a move to a sheet aluminium monocoque would be 40–60% more expensive, based on experience with the front end. For a plastic composite structure, the cost would be around 20–30% higher (Renault, 1995).

On average, the material cost of composite materials is around three to four times higher than steel, according to Renault, while aluminium costs four to seven times more than steel. The price of SMC per kilo can be compared with

Table 8.12 Material Costs (US$/lb)

Material	Cost
Steel sheet (structural)	0.75
Steel sheet (appearance)	0.75
Al sheet (structural)	3.00
Al sheet (appearance)	3.50
RTM resin	2.50
RTM glass mat	3.68
SMC	1.75
Thermoplastic resin	4.00
RRIM	2.30

Source: Dieffenbach et al. (1993).

aluminium, according to Matra, although in the US it tends to be cheaper, due to the lower oil price (see Table 8.12). However, the tools used for plastic composites tend to be significantly cheaper.

According to Matra Automobile, Europe's largest producer of plastic-intensive cars, the limit for an economical use of RTM and SMC lies between 200 and 500 a day. This takes the maximum up to 110 000 a year, Matra's capacity for the 1996 generation Espace. However, Matra points out that these volumes actually make this technology viable for quite a few niche cars produced in Europe. In France, for example, the Renault Safrane is built at a rate of only 250 a day, while the Peugeot 605 is built at a rate of only 100 a day.

In the US, plastic panels are made in much higher volumes. The GM people-carriers' production uses as many as 600–700 units of each SMC panel per day. Even at only one mould per part, a production rate of more than 400 cars a day is possible. According to research carried out by Dow Chemicals in 1987, all plastics are more cost-effective for body panels at volumes up to about 80 000 a year, after which steel can compete with thermoplastics, although RIM materials remain cheaper than steel on a per part basis beyond 200 000 a year.

Another advantage of plastics is the lower baking temperature used in the paint ovens. Although this is done in order to protect the material, significant energy savings can be made by running paint ovens at lower temperatures. However, this would not allow the combination of steel or aluminium panels with plastics, and would imply a wholesale move to plastic body panels. Where this has been done, such as at ERF trucks or Matra, or the Pontiac Fiero plant, the paint shop is surprisingly compact and involves very much lower investment levels than for steel.

The paint shop at an assembly plant has two major disadvantages:

- It is the most expensive part of the plant, on average around £200–300 million.
- It is the main source of harmful emissions in the form of VOCs into the air and paint sludge into the water.

These may in the future become more and more important considerations, and may make a move to plastics for a greenfield development particularly attractive. In fact, many plastic and composite processes can dispense with the paint shop altogether. Colour can be added at the gel coat level in many GRP and RTM processes, or it can be injected into the mould during the curing process. In thermoplastics, the material itself can be coloured; in this case the colour goes all the way through the panel, and surface damage will not alter the colour. GEP have also demonstrated a printing process for thermoplastic panels. This allows all sorts of patterns to be printed on the panel, allowing surface effects hitherto limited to custom paint jobs.

However, so far plastics have only played a minor role in automotive body panel applications. In the US, for example, in 1994 plastics took only 5% of the body panel market, although aluminium took only 3%; the remainder was steel. Like aluminium, plastics/composites suffer from the investment inertia of the motor industry which has invested vast amounts of money in steel-related technologies such as press-shops, steel body-build and paint shops. This investment cannot just be written off. Like aluminium, therefore, plastics and composites have to compete on steel's terms and that does not always show them in the best light. However, changes in external pressures on the industry such as weight reduction, lower volumes per body variant, etc., are beginning to turn the tide more in their favour.

An interesting recent example of what can be done if a vehicle is designed from first principles in plastics is the PSA Tulip experimental urban electric vehicle for multiple-user application. This vehicle features a true monocoque bodyshell (i.e. there is no separation between structural and 'outer' or cosmetic panels) made of polyurethane foam reinforced structures of glassfibre-reinforced injected polyester resin (SRIM). The level of consolidation is particularly interesting as the whole shell consists of only five parts (see also Chapter 6).

The whole question as to which body material to choose is strongly volume related and, within the plastics and composites types, General Electric Plastics have come up with the estimates shown in Table 8.13. Table 8.14 presents figures derived from Matra and Lotus sources.

Table 8.13 Viable Volumes for Alternative Materials

Material	Annual production
GRP	<1 000
RTM	500–5 000
Thermoforming	<10 000
SRIM	3 000–100 000
SMC	3 000–100 000 +
Injection moulding	3 000–100 000 +
Aluminium stamping	30 000 +
Steel stamping	30 000 +

Source: adapted from GEP.

A study by IBIS Associates from 1993 breaks this down further (Dieffenbach et al., 1993). Its weakness is that it assumes a number of fixed structures; among them a composite monocoque, which is a very rare technology at present. However, it is clear from their calculations (which apply to the US market) that in terms of fabrication costs, an aluminium spaceframe structure with SMC panels can be just about the most cost-effective car body structure up to volumes of around 400 000, beyond which steel becomes competitive again. Assembly costs for the aluminium spaceframe are relatively high, however, although still fairly close to the competing technologies. The aluminium spaceframe also does well in terms of life-cycle costs.

'Advanced composite' structures have been advanced as one means toward 'ultralight' vehicles, notably by Lovins of the Rocky Mountain Institute as part of their 'hypercar' for future vehicle technologies (see Chapter 7). The hypercar body would be a monocoque of advanced polymer composites, made from perhaps 20 components adhesively joined. In an attempt to put costs on a future advanced composite vehicle (which would probably embody carbon fibres), the Rocky Mountain Institute assumed a body mass (including doors, etc.) of 190 kg — a conservative estimate based upon an analysis of the GM Ultralite concept vehicle (see Chapter 6). The analysis required a great many assumptions to be made, for example, in terms of the availability of a robust, mass-production-ready adhesive technology. The findings are shown in Table 8.15.

If carbon-fibre prices should fall, or greater mass reductions be realised in the carbon-fibre composite structure, then the costs would more closely approach those of the steel unibody base vehicle. The above analysis is based on a volume of 100 000 vehicles per annum, and essentially puts the composite monocoque into competition on steel material terms, although as discussed in Chapter 7, there is no real need for carbon fibre to compete on price with steel in this way.

The real weakness of the ultralight vehicle concept is not the performance of the material, although it has certain characteristics which are undesirable, such as brittle failure. The weakness lies in the scale of available production technology and the overall lack of expertise in this area. A great deal of additional R&D effort will be required if such concepts are to reach the market. There is a feeling, generally, that in future the balance of investment will move towards a relative reduction in manufacturing investments, but an attendant increase in research and development (Baldensperger, 1996). A move towards a new EOV paradigm could certainly have such an effect.

8.7 SUPPLIERS, ALTERNATIVE TECHNOLOGIES AND MODULES

We have concentrated our analysis on the implications for vehicle manufacturers of a radical change in automotive technologies. It is clear that there are profound consequences for the entire automotive supply infrastructure, and indeed that the

Table 8.14 Thermoset Panel Production Processes

Process	Handlay	RTM	SMC	RRIM
Application	1. Outer/inner 2. Panels 3. Chassis components	1. Outer panels 2. Inner panels 3. Impact structures	1. Proto + low volume 2. Volume inner 3. Chassis members	Bumpers, wings, sills
Structural/non-structural	Both	Both	1. Non-structural 2. Semi-structural 3. Structural	Non-structural Impact areas
Vol^m/YR*	→ 600	→ 10 000	→ 100 000	→ 10 000

Process configuration	1. Handlay up
	   Spray up
	2, 3. Vacuum bag
	   Autoclave
		Vacuum tools (VARI)
		Clamped tools
		Press tools
			1. Vacuum moulding, or
			2. Press
			3. Press
				1. Short fibre (RRIM)
				2. Long fibre (SRIM)
				   — using preforms

Assembly technique	Rivet/bolt Bond also moulded in fasteners, etc.	Rivet/bolt Bond also moulded in fasteners, etc.	Rivet/bolt Bond also moulded in ribs + bosses	Rivet/bolt Bond foam-in brackets Snap fit
Tooling type	(a) Epoxy (b) Nickel shell	(a) Epoxy (b) Ni shell (c) Al (d) Steel	(a) Epoxy for 1 or proto 2 + 3 (b) Steel	(a) Epoxy (b) Ni shell (c) Aluminium (d) Kirksite (e) Steel
Tooling cost (£K)	(a) 1–5 (b) 5–15	(a) 10–30 (b) 20–60 (c) 15–50 (d) 50–130	(a) 5–15 (b) 40–200	(a) 15–50 (b) 30–70 (c) 20–60 (d) 20–60 (e) 60–150

Tool duration (no. off)	(a) 3000
	(b) unlimited
		(a) 2000
		(b) 100 000+
		(c) 50 000+
		(d) 250 000
			(a) LPMC 1000
			  SMC/HMC 10
			(b) 100 000 refurbish
			  every 10 000
				(a) 2000
				(b) 100 000
				(c) 50 000
				(d) 50 000
				(e) 500 000

Typical materials	Fibreglass Kevlar carbon	Glass Kevlar carbon	Fibreglass	Milled glass
Resin	Polyester Vinyl ester Epoxy	Polyester Acrylic Vinyl ester Epoxy	Vinyl ester Polyester	Polyurethane Nylon PCPD

Table 8.15 Life-cycle Economic Costs of the Steel Unibody and Composite Monocoque in Soft and Hard Tooling ($)

	Steel unibody	Composite monocoque, soft tooling	Composite monocoque, hard tooling
Manufacturing			
Material cost	353	1753	1753
Labour cost	259	240	510
Equipment cost	423	110	161
Tooling cost	325	172	271
Other cost	395	208	333
Subtotal	1755	2483	3029
Use			
Total operation	593	264	273
Disposal	(7)	(29)	(29)
Life-cycle	2341	2719	3246

Source: Mascarin et al. (1995).

character of that infrastructure will play a decisive part in shaping the emergence of new technologies. We have discussed some of the supply-side implications for materials above. Here, a more comprehensive view of the entire supply chain is given. Table 8.16 summarises the likely impact of a transition to EOV technology on the supply chain. The most obvious candidates for major change are the engine and body materials and equipment suppliers, but in practice all areas will be affected in some way.

The development of alternative technologies will take place alongside the continued restructuring of the supply base. In effect, there will be a competitive battle fought over the definition of a 'module' or system which will see all sorts of innovative combinations. That is, the status of being a module or first-tier supplier is not fixed. An illustrative example of this process is the joint venture created in 1996 between Siemens (a leading supplier of vehicle electronic control systems) and Sommer Alibert (a leading supplier of interior trim plastic mouldings) with a view to being the leading supplier of the interior module. The question of module leadership is also not fixed. For example, some vehicle manufacturers are now considering the front of the vehicle to be a 'module'. Who should control this? It could be the supplier of lights, radiators, bumpers or even front suspension. Should the front be considered a module? On the Ford Ka, the corners of the vehicle were defined as modules (sometimes called clusters).

So, while it is clear that some technologies will be redundant and that some suppliers are more exposed to these technologies than others, it is by no means clear what the shape of the future automotive components industry will be.

Table 8.16 Components and Materials in the Traditional and the EOV Supply Chain — Some Illustrative Examples

Component	Threat	Likely developments
Cylinder heads	Redundant technology	Phased out except hybrids
Ignition systems	Redundant technology	Phased out except hybrids
Radiators	Switch to plastics	Emerging role in heat management systems
Air-conditioning	Energy-intensive	Integrated into heat management systems; reduced need if glass technology improves
Glass	Lightweight plastics	'Intelligent' glass with active heat management
Tyres	Recycling problems	Low rolling resistance tyres; new approaches to recycling
Wiring	Weight	Switch from copper to optical systems; integration of functions and multiplexing
Power steering	Energy-intensive	May be redundant on lightweight vehicles + move to electric
Brakes	Weight, wasted energy	Metal matrix rotors, regenerative braking
Exhausts	Redundant technology	Phased out except hybrids
Gaskets	Redundant technology	Phased out except hybrids
Paints	Cost, economic and environmental	In-mould painting
Seats	Weight	Magnesium frames, or structural seats incorporated in body or interior module
Interior trim	Weight, recycling	Reduced use, switch to sustainable materials
Electric motors	Weight, energy consumption	Compact motors, running at higher rpm

8.8   IMPLICATIONS FOR RETAIL AND DISTRIBUTION, AFTERMARKET AND REPAIR

More and more manufacturers adopt the cost recovery route in some fashion as it is more and more difficult to make money by producing basic cars. Manufacturers make their money on the range of add-ons they offer as optional extras. In these items, the difference between the purchase price from the suppliers and the price charged to the car buyer can be as much as 400%. Some commentators therefore argue (CCFA, 1994: 9) that the market in the developed countries has reached saturation point in terms of quantity, but that there is still considerable growth potential in terms of quality. They mean by this that cars will become 'richer' in terms of features. This is increasingly where manufacturers make their money. However, although some new car buyers are prepared to specify such items and pay for them, in the used-car market most of

Table 8.17 Percentage of Value Retained After
Typical Two-Year Period (UK Market)

Item	Value retained (%)
Metallic paint	100
Air-conditioning	75
Alloy wheels	55
Automatic transmission	50
Leather seats	45
Sunroof	35
CD player	30
ABS	25
Traction control	20
Passenger airbag	10

Source: Anon. (1996b).

these features quickly lose their appeal and depreciate rapidly as a result. This is
illustrated in Table 8.17.

The future for the industry may not lie in this direction, therefore, as the used-
car market increasingly plays a leading role. This is illustrated in mature markets
by the fact that around twice as many cars each year are sold second-hand. In
France, for example, for every new car sold in 1995, 2.14 used cars were sold
(L'Argus, 1996: 100). By the mid-1990s, the total value of the used-car market
began to overtake that of the new-car market (*Fleet News*, 1994), and the
increasing importance to the market of used cars is further illustrated by
phenomena such as the 'nearly new' market in the UK (Wells et al., 1997). For the
time being, however, a prestige badge can still command a premium, allowing the
specialists at least some cost recovery. In the longer term this strategy is equally
unlikely to succeed.

All cars produced are sold through franchised sales outlets or by the
manufacturers themselves. In general terms, the sales and distribution of vehicles
account for about 35% of the total cost of a vehicle, and is now under scrutiny by
vehicle manufacturers looking for reduced cost and added competitive advantage
(Rhys et al., 1995). Vehicle manufacturers tend to have large numbers of
franchised dealers to sell and support their cars. At present, these networks are
based on the need to sell large numbers of vehicles, whereas our EOV scenario
envisages much lower volume sales.

The first implication is, therefore, that a large number of retail dealerships will
not be required, or that the income derived from sales of new cars alone will be
small. In addition, with increasing reliability and longer service intervals, the
income generated by the workshop is declining. Where novel materials are
involved (or indeed any element of specialised automotive technology) there is
considerable expense incurred in training and other support of franchised
dealerships, both for the sale and, more importantly, the maintenance and repair

of such vehicles. Manufacturers tend to respond by selecting only a portion of the entire network to sell specialised products — a feature which inevitably reduces the total sales of such vehicles.

The implications of this are clear. It will take considerable time and expense to ensure the entire network is able to handle cars built from aluminium or plastics, and powered by an electric or hybrid powertrain. A major concern for all manufacturers is the support of the vehicle once it is in the market. With respect to vehicle body materials, this concern has two basic dimensions: cosmetic damage and structural damage. It is not necessarily the case that steel is 'better' in these respects, but it is certainly 'tried and trusted' by consumers, vehicle-testing periodic inspection establishments and, as importantly, by insurance companies.

The insurance industry determines the base from which insurance costs are derived by looking at two issues: damageability and repairability. On the basis of their analysis, vehicles are then classified into insurance groups or categories. Novel materials and designs, simply by being different, may be placed into higher insurance groups and, as is discussed below, may force special measures to be taken to guarantee the security of structural repairs. In fact, the Renault Espace was moved up several categories by UK insurance companies a few years after its introduction in the UK, when the full repair implications of its novel body construction became apparent. In assessing vehicle accident performance, the UK insurance industry research centre at Thatcham uses a 15-kph full-front impact, 40% offset (right) front and (left) rear — the most common impacts — to assess the two issues. Most impacts over 30 kph are insurance write-offs. These are more or less internationally agreed standards.

- *Damageability*. How well does the vehicle sustain damage? As a basic rule, all vehicles should contain damage to the first 300 mm, but the performance of contemporary steel vehicles ranges widely. An important element here is the quality of the bumper system; aluminium beams are very good at absorbing impacts, and are likely to find further applications in the future.
- *Repairability*. How far back into the structure does the repair need to go? How easy to perform is the repair task? This is also a question of vehicle design. In some cars damage may travel further back, resulting in repairs carried out further into the vehicle structure. In some cases the design of the body may be such that undamaged parts have to be cut out and replaced in order to make a repair.

With steel bodies the need to repair is easily identified, and the effectiveness of the repair is easily verified by examination of the weld quality. This is not always the case with structures that have been bonded or weld-bonded.

For example, in the UK, Audi has designated only part of its network as repair centres for the Audi A8. In order to minimise the training and special equipment costs involved in preparing a franchised dealership in the repair of the A8, Audi followed a strategy which identified different levels of repair, and hence different levels of skill and equipment, at the franchised dealership. In the UK, the initial plan called for just three 'Category A' sites which would be able to carry out all types of structural and cosmetic repair.

While such an approach does reduce costs in installing a network to support vehicles with novel materials, there are clearly problems too. One of the more significant weaknesses is the question of customer care, where customers may be surprised and disappointed to learn that their vehicle has to be removed some distance for repair — inevitably there will be a time penalty. Another weakness is that franchised dealers excluded from involvement will tend to press for equal status, especially if their dealership has sold an A8 to a local customer.

In the longer term it is difficult to see how Audi can retain the A8 within its own network, despite the intentions of the insurance companies. While the cars are within the network, and any structural repairs are carried out by qualified Audi staff, it should be possible to avoid problems arising from repairs which would damage the brand reputation. However, if the vehicles do end up in the independent aftermarket repair sector, or if owners themselves attempt repairs, then there is a potential for problems to emerge.

Pressures towards a more 'sustainable' industry may force the development of energy-efficient or alternative-fuelled long-life cars, built in much lower volumes, but whose income is generated by high residual values, bought and sold many times through the dealer network, with regular retrofitting of the latest and cleanest driveline technologies. These high-value, high-cost products could remain the property of the manufacturer or dealer who would lease them to customers, thus making the initial cost less important. This would lead to a much closer link between dealer and manufacturer, as only in this way could the manufacturer tap into this lifetime value stream. Manufacturers who owned their networks would clearly be at an advantage.

These developments could blur or 'fuse' the current distinctions between new sales, used sales, service and repair/maintenance into one parc-based revenue stream which should be more predictable, less vulnerable to short-term market fluctuations and which would offer repeated opportunities for improved customer contact, repeat business and customer loyalty generally.

9 The Motor Industry Beyond 2020 and the Death of Motoring?

... in the automobile industry, 30% of the parts are made up of electrical and electronic components. It is predicted that within 10 years, this ratio will increase to 50%. In the year 2010, this figure could be as high as 60% or 70%. Then it will be difficult to determine whether it is an automobile industry or an electrical/electronics industry. As the sectors disappear, the design factor will play a bigger role in the automobile industry ... Advanced industries in the 21st century will be comprised of multiple industries ... [the] automobile and electronics industries incorporate various aspects of other industries into their research and product development. Survival in the 21st century will not be possible if we continue to separate individual industries.

(Lee, Kun-Hee, Chairman Samsung, in Lee, 1994: 98)

9.1 INTRODUCTION

In previous chapters we have pointed the way very clearly and unambiguously to a period of radical change in the motor industry. However, the future does not need to be as uncertain as this would suggest. The future does not just come about; it is something we make today. Decisions made by the industry and by the regulators today will bring about that future. In order to make it happen we have to paint a clear vision of this future, something we have been trying to do with the EOV concept. If everyone in the industry, in government and in the wider community begins to share a certain vision of the future, incorporating the ideas presented in this book and elsewhere, then change is possible and it will be able to benefit as many people as possible. As Ealey and Gentile (1995: 107) put it: 'Companies that begin thinking about these issues now, and formulating approaches on how they could address them in the future, will have a significant advantage over those companies that allow their responses to evolve haphazardly.'

However, Cronk (1995: 6) also points out that: 'Dramatic changes such as these are not managed well by traditional "Fordian" organizations, or even by the more modern "lean production" organizations.' The reasons for this are clear, as both 'systems' have come to imply the use of the Buddist manufacturing system, with its attendant high fixed costs, consequent high production volumes and low profitability. Raff (1994: 1) argues that 'no system that fails to turn a reasonable profit will persist'. However, some of the existing players may yet be able to adapt to the changing circumstances. In terms of strategic posture towards environmental issues, firms may occupy a position along a continuum of alternatives which range from outright opposition to pro-active. Norcia et al. (1994) identify four main positions a firm may take: resist, comply, accommodate and pre-empt. The 'resist' position reflects a negative and short-term defensive attitude to environmental issues. At the other extreme, the pre-empt position reflects a strategic vision of pro-active management seeking to lead and control the development of environmental issues.

A similar idea is expressed by Trisoglio (1993) who identifies three management philosophies to respond to environment-driven change: environmental management, environmental competition and sustainable development. The 'environmental management' position in the Trisoglio framework reflects a defensive 'environmental quality as cost' attitude via overt government control leading to 'end of tailpipe' technical solutions. Under 'environmental competition' a more positive stance is entailed; improving environmental quality is seen as a business opportunity, while a broader life-cycle view is taken of environmental issues. With 'sustainable development' the environment is central to corporate activity, with strategy designed around and based within ecosystem tolerance.

For firms in the automotive industry, the current level of strategic response suggests that incrementalism is the preferred strategy, reflecting a concern to manage, control and shape environment-driven change. It has been claimed by some that capitalism in general, and the automotive industry in particular, has moved into a new era of knowledge-intensive production. Florida and Kenny (1993: 642) state:

> Increasingly, automobile and steel companies must produce software, integrated circuits, programmable logic controllers, advanced robotics and machine tools and pieces of equipment. These companies are developing their own software capabilities, spinning out software companies, and investing in high-technology start-ups as they endeavour to compete in the new age of technology-intensive, digitally based manufacturing.

This, in terms of the environmentalist/ecological paradigms noted above, hardly represents a new epoch as the authors claim. Rather, it represents a continuation and refinement of the ability of the industry to respond in a technical sense. Indeed, Florida and Kenny may be right in that if the incrementalist position is maintained, the industry will become more like the model they outline. But they fail completely to understand the scale and systemic nature of the crisis facing the automotive industry as a whole. Equally, they are probably right to suggest that innovation at all levels is the key to transformation and survival, but have failed to address the impact of this feature on the core competencies of leading firms. Here, Topolansky (1993: 47) concludes that:

> ... international competition is likely to overcome resistance to change, and propel the global automotive industry towards a new generation of technology ... those

automakers that have failed to keep up with the pace of progress may find themselves facing severely depleted market shares, significant inventory backlogs and ... a fatally damaged international reputation.

In this case, a more radical transformation may be required in the form of a paradigm shift. In this final chapter we consider first the ability of the automotive industry to adapt to change in historical perspective. We then move on to consider a range of related issues concerning the future of the industry, including the scope of 'traditional' restructuring to sustain the existing industry; the means by which transitional, interim or partial changes may be introduced, and the global competitive context within which these changes are likely to occur.

It is not our purpose here to translate this thinking into concrete predictions on, for example, the numbers of electric vehicles sold by the year 2005. Neither do we seek to speculate as to exactly which firms might survive or fail and why. Indeed, a feature of the immediate future for the automotive industry is that it will be volatile, less predictable and not amenable to the traditional method of forecasting via the extrapolation of existing trends.

9.2   SECTOR SHIFT AND STRUCTURAL CHANGE IN HISTORICAL PERSPECTIVE

Clearly, some companies are going to be better able to adapt to these dramatic changes than others. Those manufacturers with a wider product portfolio beyond cars may well be at an advantage as they are more likely to be able to shift their core product areas away from cars or away from existing car-making technologies. Similarly, manufacturers with a recent history in other product areas may be more flexible than those firms that have only ever been in the car business, although the latter may be more committed to finding radical solutions in order to stay in the industry. At the same time, manufacturers are increasingly passing on responsibilities to their suppliers. In the MCC Mercedes-Swatch production system for the Smart car, MCC limits its activities to the integration of a number of modules pre-assembled in separate units by major systems suppliers. As technologies change, manufacturers may have to rely more on suppliers who have greater expertise in these technologies. Examples are ALCOA or Norsk Hydro for aluminium chassis, and DSM or General Electric Plastics for plastic body panels or the growing importance of the electronics systems in a car. Electronics may soon represent up to 40% of the value of a car.

To try and assess how change might happen, we have once again taken a look at the history of the industry in order to determine from which sectors firms were most likely to move into car making in the past. A number of sectors certainly stand out. In Europe especially, many firms moved from another recent invention, the bicycle, into car making. It is significant that all modern bikes, the so-called 'safety' bicycles, are based on the Starley safety bicycle, known as the Rover. This was the basis of the Rover car company. However, the list in Table

Table 9.1 Bicycle Makers Moving into Car Making

Adler, Belgica, Belsize, Bianchi, Brennabor, BSA, Calcott, Calthorpe, Century, Cito, Clement-Bayard, Clyde, CM (Mochet recumbent bikes), Condor, Coventry-Premier, Dansk, Darracq, Dennis, Duesenberg, Dursely-Pedersen, Eysink, Gazelle, Gladiator, Graef und Stift, Humber, Hurtu, Invicta, Kondor, Lea-Francis, Minerva, NSU, Opel, Panther, Parisienne, Peerless, Phaenomen, Pierce-Arrow, Presto, Progress, Puch, Queen & Russel, Raleigh, Rambler, Riley, Rochester, Rochet, Rover, Royal Enfield, Rudge, St Laurence, Scania, Simplex, Singer, Star, Sunbeam, Swift, Terrot, Vallee, Van Gink, Victoria, Wanderer, Warwick, Welch, Wilton, Winton, Vivinus-Orion

Source: data from Georgano (1968).

9.1, which is by no means exhaustive, shows a large number of other historic and existing vehicle makers that made bicycles before they moved into cars. Alternatively, they are bicycle makers that have had a brief car-making venture.

Many of these also made motorbikes, while some moved into car making via motorbikes, such as DKW, Imperia of Belgium, Brough, Imperia of Germany, Innocenti (Lambretta), JAWA, Laurin & Klement (Skoda), Magnet, Maico, Matchless, Rex, Rovin, Rudge, Thomas and Triumph.

The US equivalent of bicycle making seems to have been the building of horse-drawn buggies and many US firms moved from there to cars. However, it is interesting that three of the top US luxury car makers — Peerless, Pierce-Arrow and Duesenberg — all started with bicycles. Other sectors that stand out as ones from which to enter the motor industry were guns and arms, coachbuilding and carriage making, railway rolling stock and locomotives, and aircraft (Table 9.2).

Other sectors feature less prominently, such as aero-engines (Salmson, BMW, Anzani, CMN) and agricultural machinery (Glas, Lamborghini, Case, International, Kissel, Piccolo, Ruston & Hornsby). A number of firms also moved from being suppliers to the motor industry, although in a sense they were already part of it. Several suppliers of proprietary engines, for example, took up car making (Ballot, Anzani, Aster), while Willys-Overland started by making wheels. Other sectors such as ironworking (Stoewer, Thames) or diecasting (Franklin) also appear somehow related. However, others seem to have moved from seemingly unrelated sectors, such as woodworking machinery (Panhard, Vermorel), photographic plates (Stanley), candle making (Lanza), bobsleighs (David), sheep-shearing equipment (Wolseley), sewing machines and textile machinery (Toyota, Saurer, Suzuki, Weber), shipbuilding (Euskalduna, Kromhout, Omnia, Daewoo, Hyundai), shipping (Lloyd) and electric cranes (Royce). Electric and domestic appliance makers Miele, Westinghouse, AEG and Siemens have all made cars also.

We have seen that five sectors in particular stand out: bicycle making, armaments, coachbuilding, railway rolling stock and aircraft. Four of these can be considered related transport areas, while guns and armaments was an early sector to adopt precision engineering and series production techniques, which were particularly suited to the emerging car industry. Besides, arms and aircraft

Table 9.2   Firms Moving into Car Making from Other Sectors

Railway	Arms	Coachbuilding	Aircraft
Vabis	FN	Auburn	Antoinette
Tatra	Hotchkiss	Brewster	Bleriot
Decauville	Ansaldo	Dort	Bond
De Dietrich	BSA	Facel Vega	Breguet
Delaunay-Belleville	Royal Enfield	Holden	Bristol
Diatto	Husqvarna	Jensen	Burney
Hanomag	Skoda	Lohner	CEMSA
Raba	Martini	Nesselsdorf	Davis
Metallurgique	Mauser	Rauch & Lang	Ellehammer
	Nagant	Sayers	Farman
	Steyr	Spyker	Heinkel
	Z	Stirling	Messerschmitt
		Studebaker	Otto
		Szawe	Piaggio
		Velie	Prince
			Rumpler
			Saab
			Thulin
			Voisin

Source: various.

are sectors where demand fluctuates violently, with demand peaks at times of war. Many firms wanted a more secure basis, which they sought in car making.

Coachbuilding was clearly a related activity. As the horse was phased out and the car emerged, the carriage makers either closed their doors or moved into the new industry, while many railway suppliers clearly saw the motorcar as both related and emerging as a possible competitor. The type of technical expertise of these industries was also clearly relevant to car making. The same also holds for bicycles which, as we saw in Chapter 4, often shared basic techniques with early cars, especially in the area of bent tubes for frames and chassis. How pervasive such expertise can be is illustrated by Rigby (1996). He cites the Greek writer Pytheas who visited Britain in 330 BC and who praised the British chariot industry. Britain also became an important carriage maker, invented the safety bicycle and became a major force in car making. Even today, UK expertise in product design, high-technology vehicle development and racing car design is unmatched anywhere in the world.

Another feature which applies particularly to the bicycle industry is that this was a new, dynamic and entrepreneurial industry at the time. It supplied playthings for a wealthy clientele, subject to passing fashions. The character of Mr Toad in Kenneth Grahame's *Wind in the Willows* satirised this type of behaviour at the time. However, the essence of a young type of dynamic agile firm entering the car industry may also be applicable to the future scenario we are painting, i.e. that of a radically new type of car industry.

Even the railways were still relatively new when the car industry was in its early years, and subject to rapid technical developments. Hence, the railway rolling stock and locomotive makers are not a surprising source of early car-making entrants either. Again, today, we can see new developments in high-speed rail rolling stock which may be relevant to the EOV type of car making. The new high-capacity TGV Duplex rolling stock, for example, incorporates large amounts of aluminium and magnesium in order to limit weight increases to within the maximum 17 tonne axle weight. It is reported that the fact that aluminium needs only one coat of paint alone saves some 300 kg per carriage. Future versions are likely to be made of composite materials in order to reduce weight yet further (Leniger, 1996: 23). Although the material cost is around four times higher than the aluminium–magnesium option, SNCF expect the overall cost to be lower as a result of reduced maintenance requirements, absence of corrosion and cheaper production methods (Leniger, 1996: 24). Clearly, there are lessons here for car making.

This theme of the source of likely future entrants to the car industry is also developed by Cronk (1995), who argues that the E-Motive industry will be built by a specific type of company. New entrants could be drawn from several contemporary and dynamic sectors with relevant technological expertise. The electronics, computers and information technology, military and aerospace industries could yield future car makers in an EOV scenario, as could small firms with composite expertise, such as those in boatbuilding, surf-boards and materials suppliers, for example.

It is also relevant to consider some of the exits from car making. This was clearly easier in the days of craft production than it is today. There are quite a few examples of companies that moved from making both cars and trucks to just concentrating on trucks, thus giving up car making, though not leaving the industry. However, it does avoid many of the investments needed for mass production of cars. Examples of firms that took this route are Albion, Autocar of the US, Berliet and Unic of France, Berna in Switzerland, Miesse in Belgium and Scania-Vabis. Another group of companies moved through a car-making phase, after which they either returned to their original core business or moved into an entirely new sector. We list a few examples of this in Table 9.3.

The diffusion of mass production

In the past there has been another major paradigm shift in the car industry. This was the shift to mass production and the Budd paradigm from the previous craft production phase of building a mixed-material body on a steel chassis. This shift took some years to complete, starting with the changes in machining and assembly initiated by Ford, then combined with the changes in body/chassis technology initiated by Budd. This process took from around 1915 until the 1950s to complete; in fact, refinements are still being made. However, the main push came in the 1920s and 1930s, as we saw in Chapter 4.

Table 9.3　Firms That Have Entered and Exited the Car Industry

Firm	Sectors
Hotchkiss	Arms → cars, trucks → military vehicles
FN	Guns → bicycles, motorbikes, cars, trucks → guns, aero-engines
Albion	Cars and trucks → trucks → axles, gears, camshafts
Amedee Bollee	Bell foundry → cars → piston rings
Bean	Car components → cars → car components
Hispano-Suiza	Cars → cars, aero-engines → aero-engines (SNECMA)
Mieussel	Fire extinguishers → cars → fire extinguishers
AEG	Electrical → cars (NAG) → electrical
Peerless	Clothes wringers, bicycles → cars → brewing (Carling)

Nevertheless, Chapter 4 painted perhaps a partial picture. We focused there on the companies that made the shift from craft to mass production. However, as Raff (1991, 1994) points out, the shift was by no means universal. He describes (Raff, 1991) how in the US, the first to adopt true mass-production methods after Ford was in fact the Chevrolet division of GM. This happened in the late 1920s, so a dozen years after Ford and a dozen years after the first implementations of Budd technology. Other GM divisions followed over the next decade. Most of the US car industry remained craft-based, for, as Raff (1991: 727) explains, 'the key to Ford's production success was not understood by contemporaries'.

At that time, as in popular thinking today, the key to Ford's success was thought to be the moving production line, but Raff (1991: 731) points out that:

> ... the central innovation at Ford was not the conveyor belt ... The real source of the huge Ford output lay in pushing the limits of the American System by using large-scale investment in the requisite machines and tools to do away with most direct production tasks that required discretion. And that does not seem to have been common.

In fact, most US car makers at that time did not adopt mass production at all, and most US car makers also gradually went out of business. Raff (1991: 731) argues that the diffusion of the mass-production paradigm through the US car industry 'went on principally through the entry and exit of firms rather than through change in the methods of ongoing firms'.

In 1929, the Big 3 were responsible for only about three-quarters of the output of the US auto industry, while they operated only a quarter of the factories (Raff, 1994: 12). Most of the remainder were makers of quality upmarket cars, such as Pierce-Arrow, Packard, Duesenberg, Cord and others. Raff (1994: 15) explains that, contrary to popular belief, this part of the industry was not wiped out because its average cost curve was too high. In fact, these companies could and did practise cost recovery and had better margins than the mass producers.

Raff argues that the reason these firms failed was that, unlike the mass producers who had sunk assets they could dispose of in the Depression, these firms' assets were unsunk in that they consisted of the skills of their workforce. Redundancies meant loss of skill and expertise. In trying to retain their skilled workers they ran out of time. The Depression lasted longer than expected and these companies were forced out of business. They simply ran out of money to retain their principal asset. In Europe, too, the artisanal quality producers were the ones who suffered for the same reasons. The case of France is illustrative.

France had a number of 'quality' (i.e. non-mass production) car makers between the wars, notably Delage, Delahaye, Hotchkiss, Voisin, Hispano-Suiza, Talbot, Bugatti and Salmson. The 1930s saw some restructuring as Voisin was first taken over by a Belgian group trying to build a European equivalent to GM, and then closed down in 1939. Hispano-Suiza pulled out of car making in France after 1937 to concentrate on its aero-engine interests. Delahaye took over Delage in the 1930s and was then taken over by Hotchkiss after World War II. What little survived after the war was essentially killed off by a combination of punitive taxation on large cars and the advent of the mass-produced Jaguar luxury cars in the 1950s. Bugatti did not really survive the war, although the remains of the company were taken on by Hispano-Suiza, now SNECMA, in 1961, while Salmson went to Renault and Talbot to Simca in the late 1950s. The gradual decline is illustrated in Table 9.4.

Some of the failing companies in both Europe and the US managed to retain some form of presence until the late 1950s. This is some 30 years after the introduction of the new paradigm. In Europe, too, we saw the survival of non-mass, non-Budd firms for a long time. In fact, we pointed out earlier that quite a few non-Budd craft producers survive today and, in some cases (such as Morgan and TVR), are quite profitable. Some others are retained by large mass producers and are thus able to survive, e.g. Ferrari and Aston Martin. These two firms use hand-beaten aluminium bodies on separate chassis, a very labour-intensive method particularly suited to expensive cars, although Morgan has made this work on much cheaper cars by retaining an ash frame (Figure 9.1).

We saw earlier that the very fact that they had not adopted the Budd paradigm has allowed some niche players to survive on low volumes. The lower investments, product development and tooling costs are the key to their ability to survive. A similar picture may not arise for the next paradigm shift as low

Table 9.4 Car Sales in France: Selected Companies and Years

Company	1939	1950	1955	1959
Citroën	54 240	38 434	106 163	158 815
Renault	33 462	55 080	125 439	169 119
Hotchkiss	1 935	1 350	140	1
Salmson	852	960	87	—
Talbot	531	290	13	6

Source: adapted from CCFA.

Figure 9.1   The Morgan Timber Body-frame on a Steel Chassis made Profitably in Low Volumes

volumes are unsustainable in the Buddist paradigm. The next change may therefore be more sudden. On the other hand, the supposedly weak players, i.e. those most committed to the current paradigm, are also the ones with the greatest resources. From this point of view their survival would seem more certain, at least if they adapt. In fact, they are showing some signs of adapting. GM's EV1 is a case in point.

9.3   THE VEHICLE MANUFACTURERS AND THE SCOPE FOR PARADIGM SHIFT

Clearly, on past performance we cannot assume that the current players in the industry represent a stable situation. It is possible for firms to enter and leave any sector. The only problem is that the economics of car making at present represent a particularly high barrier to both entry and exit. However, consider a new situation, based on the EOV paradigm. This brings with it a new set of economics which are likely to make it easier for new entrants to come into the industry. At the same time, the existing players may still be to some extent burdened with the high exit costs inherent in the Buddist paradigm.

In order to assess and compare the different players' abilities in these respects, we have assembled Table 9.5, which provides a simple index-based assessment of potential for change.

We have identified a number of relevant parameters:

Table 9.5   Car Makers' Ability to Shift Core Product Areas

Firm	A	B	C	D	E	Total
BMW	2	1	0	2	6	11
Citroën	3	1	0	4	2	10
Fiat	0	0	2	4	6	12
Jaguar	1	1	0	1	0	3
Mercedes	0	0	2	4	6	12
Nedcar	1	3	0	1	2	7
Peugeot	3	1	1	4	6	15
Renault	0	0	0	4	8	12
Rover	1	1	0	2	2	6
Saab	2	2	1	1	0	6
VAG	1	1	0	4	6	12
Volvo	0	0	1	1	2	4
GM	0	0	2	4	8	14
Ford	0	0	2	4	2	8
Chrysler	0	0	0	2	6	8
Toyota	3	1	1	4	6	15
Nissan	0	0	0	4	2	6
MMC	3	1	2	4	2	12
Honda	1	3	1	2	2	9
Mazda	0	0	0	2	2	4
Suzuki	1	3	1	0	2	7
Subaru	0	0	1	1	2	4
Daihatsu	1	3	0	1	2	7
Hyundai	2	4	2	2	2	12
Kia	1	2	0	2	2	7
Daewoo	3	4	2	2	2	13
SsangYong	3	4	2	0	0	9

A   past shift/expansion from other sector
B   timing of shift
C   current level of diversification
D   resource base
E   willingness to change and wider political context

Within category A we recognise three sub-categories, and the points score increases the further removed from car making the shift is, as follows:

- shift from other wheeled transport (trucks, motor bikes): 1 point
- shift from other transport (ships, aircraft): 2 points
- shift from non-transport area: 3 points

Under B we determine the timing of the previous sector change. The longer ago this took place, the less likely the relevant experience is still to be in existence within the organisation; the benefits therefore fade over time. We have quantified this as follows:

- shift in past 20 years: 4 points
- shift between 20 and 40 years ago: 3 points
- shift between 40 and 60 years ago: 2 points
- shift more than 60 years ago: 1 point

Although few current staff are likely to have experience of a shift that occurred over 60 years ago, it is still part of company history and culture, and can still be counted. Where more than one applies on categories A and B, the highest score is used. The next category is C, where we give a score for the current level of diversification relative to other car makers, as follows:

- very diversified: 2 points
- semi-diversified: 1 point
- automotive and related only: 0 points

In category D we essentially distinguish small, medium and large size companies. This rating combines production capacity/volumes, employees, turnover and profits, to give four ratings:

- large: 4 points
- medium-sized: 2 points
- small but linked with a big partner: 1 point
- small: 0 points

Where brands are part of a larger company such as BMW-Rover or PSA (Peugeot-Citroën), each brand receives the rating appropriate for the group. For a joint venture company such as Saab (50% GM) or NedCar, some allowance is made for their own size and that of the dominant partner, from whose dominance they benefit in some ways.

Under category E we have a dimension which reflects the current ability and willingness of a company to move into alternative vehicle technologies based on recent experience in this area:

- more than one alternative technology model in the range in the 1990s: 8 points
- one alternative technology model in the range in the 1990s: 6 points
- concept car embodying alternative technology in the 1990s: 2 points
- none of the above: 0 points

All these then add up to a total score for each company, which gives an assessment of the likely ability to shift core product areas, i.e. the likely flexibility in terms of moving main product areas if necessary, but also flexibility in moving to alternative technologies for car making. Of course, we recognise that this index is arbitrary and simplistic; a detailed analysis would have to give far greater consideration to the unique circumstances of each company, including the political context within which it operates. Quite simply, it is clear that no national government is likely to allow a major car company to collapse totally; intervention can and will be used to slow the pace of change or ease the corporate costs of restructuring. Moreover, as was illustrated in Chapter 3, the social mobilisation of technological resources at national and international level will play an important part in the fortunes of individual firms. However, the index is useful for an 'at a glance' picture of the potential within the existing car companies to respond to the challenges we see emerging.

Table 9.5 further illustrates how diverse, at the company level, the automotive industry is, and thus that the base from which a transition could be undertaken varies widely. Table 9.5 shows a rather stark picture in that it polarises these issues. However, in doing so it shows more clearly which firms are prepared for change and which are less prepared. In general terms we could say that those firms scoring at least 10 have a reasonable chance of riding the storm, either through their size or their adaptability, although the real world is not usually that simple.

It is interesting, for example, that despite our view that regions outside the West are not well placed to be the source of radical change in the automotive industry, it is companies from Japan and Korea that score highly. The nature of Korean car firms is an important factor here. In Korea, cars are made on the whole by the chaebol, the large diversified conglomerates that dominate Korean industry. Apart from Kia, all are members of a chaebol, while the largest, Samsung, is about to enter car making. It is clear that these groups have the option to pull out of car making, if necessary, as they can rely on their other activities to pull them through.

Given that firms exist within specific nations or regions, the position of each firm will in effect be 'filtered' by the social and economic context within which they operate. Table 9.5 groups the companies by region or country. In Europe there is a broad spread in terms of ability to change, from Volvo (as one of the most susceptible to change) to Peugeot, part of the PSA group (as the most likely to cope with change). In North America, GM and Ford have both the scale and the expertise to survive, but the position for Chrysler is more precarious. Of the Japanese companies, Mitsubishi and Toyota stand out, while in Korea it is Hyundai and Daewoo. Co-operative ventures between car makers can also help in improving the prospects of individual firms. The government-sponsored collaborative projects such as the PNGV, outlined in Chapter 3, are a case in point. These specifically shield individual firms from the worst effects of a possible paradigm change.

9.4   TRANSITION IN A GLOBAL COMPETITIVE CONTEXT

The normal process of competition will not be suspended as the industry makes its way towards the paradigm shift. Rather, competition is inseparable from the processes of technological change that we envisage. There are some features of the global competitive context which are worth highlighting here: the limits of traditional restructuring, and the scope for emerging economies as producers and consumers of cars.

Limits of traditional restructuring

It might be proposed that the automotive industry could cope with the economic pressures it faces through the 'traditional' process of industrial restructuring, that of consolidation of ownership through acquisition and mergers. Indeed, many commentators have pointed to the need for such restructuring in both Japan and Europe. So, if there is an economic logic, why has consolidation not occurred? There are perhaps two main reasons:

- the politics of industrial restructuring
- the difficulty of achieving real gains

With respect to the politics of restructuring, it is the case that by virtue of their size and economic importance, the fortunes of the major automotive companies are of considerable interest to their national governments. The aborted merger between Renault and Volvo in the early 1990s clearly illustrated the sort of political problems that can beset such a proposal, even when the two chairmen of the companies concerned had reached an agreement. In the event, the merger was eventually rejected by Swedish institutional investors, who refused to allow their assets to be transferred to the still-nationalised Renault. Equally, it is politically difficult for any government to stand to one side if a major company faces collapse, though the 'non-interventionist' UK did refuse to support Leyland-DAF after the parent Dutch company became bankrupt.

Just as important, however, are the limited gains to be made from mergers and acquisitions, and how long those gains take to realise. In the case of Iveco, for example, it took about 15 years for the entire range, production structure, distribution network and supply base to be integrated into one cohesive system. Even by the early 1990s, a number of separate parts numbers systems were in use, while only the recently introduced Euro range has achieved a common cab family. Quite apart from the 'technical' aspects of achieving such integration — and integration of this type is vital if there are to be economic benefits — there are more subtle but equally important differences in corporate culture, tradition and style which can impede progress.

The most likely format of industrial restructuring for the future is the continued proliferation of partial alliances and joint ventures, and pragmatic short-term measures such as the 'badging' of competitors' products. PSA chairman Calvet expresses his views on these issues thus: 'I believe partial alliances will multiply. PSA cooperates with Renault. We sell engines to Nissan ... But I do not believe in large mergers. Excuse my language, but bringing Peugeot, Citroën and Talbot together was a pain in the ass' (Johnson & Farhi, 1996).

Emerging economies as consumers of new cars

There is less prospect that emerging economies will provide new growth in sales than many in the automotive industry appear to believe. Although even limited growth in countries with large populations such as India, Indonesia and China will add significant incremental sales, much of this will be from local production and assembly. However, these countries will never reach the same levels of motorisation of the West. Many Asian cities already suffer severe congestion at relatively low car ownership levels; Shanghai has traffic jams of bicycles. In several instances there have been 'anti-car' measures introduced.

In Thailand the aim is now to reduce the 13–14 million daily trips into Bangkok to 4 million people and 3 million vehicles (Jatoorapreuk, 1996). In addition, a mass transit system will be built for the city. In Jakarta, Indonesia, a three-in-one rule was introduced, whereby between 6:30 and 10:00 a.m. only cars with at least three occupants are allowed access. Although professional occupants have appeared who hire themselves out, the measure still acts as a deterrent, leading to a reduction in traffic during these times of 41%, of which 64% were passenger cars, and an 18% reduction in travel time (Kubota, 1996a: 17).

However, what is interesting is the extent to which people have changed their behaviour to accommodate the measures, with many businessmen now scheduling meetings from 11:00 a.m. onwards, thus leading to an overall improvement in the spread of traffic generated (Kubota, 1996a: 18). Manila in the Philippines also has a car control policy. It is based on number plates, with numbers ending in 1 or 2 banned on Mondays, those ending in 3 or 4 on Tuesdays, etc. The ban covers cars, trikes, motorbikes, and even buses and taxis, although trucks and emergency vehicles are exempt. There has been some traffic reduction effect, although multiple car ownership limits the scope somewhat (Kubota, 1996b: 25).

In the case of Singapore, punitive fiscal disincentives to car ownership have been introduced alongside measures to restrict car movements and adopt 'roadside intelligence' systems to control flows (Toh, 1992; Olszewski & Turner, 1993; Rao & Varaiya, 1994). In Singapore there is a high level of disposable income, but new cars face a 200% import duty which results in even basic entry-level models costing S$80 000–90 000 (£35–40 000). In order to purchase such a car, potential consumers have first to bid for a permit — typically costing around S$40 000–50 000 (£17–22 000). The permit lasts 10 years, but has to be sold with the car on the used market. Allied to the extensive controls exerted over individual movement in vehicles in Singapore and the very well-developed public transport infrastructure, it is not surprising that the market for new cars is extremely small.

Additionally, it is clear that in the fast-growing economies of the world and in those markets where the automotive industry believes new car sales will grow, infrastructures cannot be expanded rapidly enough. Moreover, the combination of outdated vehicles, poor roads and bridges, inadequate vehicle servicing, poverty, mixed transport modes and even cultural attitudes are leading to an appalling death rate from vehicle accidents in many of these emerging economies (Seymour, 1996). The prospect of such emerging economies all achieving contemporary western levels of car ownership, even with contemporary levels of automotive technology, is one which does not bode well from an environmental perspective, whatever economic and equity arguments there may be to support such motorisation.

The case of the Korean automotive industry

The domestic market in Korea expanded rapidly from 1989. Car ownership per head grew from 38.8 per 1000 people (1987) to 144.0 per 1000 by 1993. While not high by international standards, further rapid growth in the level of car ownership is not expected in the Korean market. The government has enacted a new 'Transportation tax' which, coming into force in 1994, has raised running costs considerably. Additionally, any household that purchases two cars will pay double the acquisition and registration tax for the second car. Increasingly, registrations will only be allowed where the car owner can prove that garage space is available for the vehicle. In the transition towards a buyers' market in Korea, the domestic producers have been forced to abandon simple price leadership strategies and accept reduced margins in the drive to increase or sustain market share. The domestic market situation and the realisation that, ultimately, the Korean market cannot continue to be protected from international competition, has underpinned the new drive for export market growth.

If all the capacity expansion plans are realised by Korean automotive firms (both inside and outside Korea itself), there will be a potential output of 5.8 million vehicles by the year 2000, as shown in Table 9.6. Allowing for kit assembly

Table 9.6 Projected Capacity Expansion in the Korean Automotive Sector to the Year 2000 (000s Units)

Firm	1994 capacity	2000 capacity
Hyundai	1150	2000
Hyundai Precision	76	—
Kia	551	1500
Asia Motors	146	—
Daewoo	396	2200
Daewoo Shipbuilding	204	—
SsangYong	39	60
Samsung	0	40
Total	2562	5800

Note: group figures. Assumes two-shift pattern and 265 working days per year. All categories of vehicle. A significant proportion of the new capacity will be outside Korea.
Sources: KERI (1994), CAIR estimates.

operations intended purely for local markets, transplant production outside Korea but not inside Europe, and for demand in Korea itself, Korean automotive firms could have about 2.5 to 3.0 million vehicles to release onto global markets by the turn of the century. Of these, perhaps 500 000 to 600 000 may be directed at Europe. While these figures are a 'worst case' scenario as far as West European industry is concerned, progress by the Koreans to date suggests that the ACEA estimates are not unreasonable. ACEA predicts that, should the Korean expansion in capacity go ahead as planned, imports into an already crowded European market could reach 360 000 units per annum by the year 2000 (ACEA, 1994c).

The Korean automotive sector is not just focused on capturing small shares in the major global markets. Internationalisation has proceeded rapidly. Rather than rely upon the North American market, the Korean firms have spread their risks, with emerging Pacific-Rim countries occupying a prominent role in their long-term strategies for expansion. The favoured locations are outside the advanced industrial countries, usually those countries lacking an indigenous capability, with an emerging market. In many cases these investments are intended to serve local markets, but have the potential to be integrated into a more global structure of production in the future. If nothing else, the expansion of the Koreans suggests that competition throughout the global market will only intensify in the future, with the possibility that they will be joined on the world market by Proton (Malaysia) and possibly firms based in India.

One of the unique strengths of Korean industry is the chaebol structure. The chaebol are the large diversified industrial conglomerates that dominate the South Korean economy. They are similar to the zaibatsu that played a key role in Japan's industrialisation, but which were forced to break up by the US occupiers after 1945. As a result, Japanese industry reformed along the lines of the keiretsu, much looser networks of independent companies. Recent entrants such as Honda tend to fall outside this system.

Most of the chaebol were formed after the Korean war, although a few date back further. Their history reflects the history of Korean industrialisation in that many of their activities were prompted by the fact that these activities, needed by the group, were not available in Korea at the time. SsangYong provides an example in that when it moved into cement production it found that it needed paper bags to pack the cement. It therefore set up a paper mill. To provide transport for its cement it moved into shipping and into electricity generation to provide the power not available at the time (MIRU, 1990: 108). The move into construction was part of downstream vertical integration from cement. The move into vehicles came only in the 1980s and allows SsangYong to use its own trucks on its construction sites. These activities are spread over 19 affiliated companies. Education interest involves Kookmin University and a number of associated institutions; the leisure activities involve a number of holiday resorts built and managed by SsangYong group companies.

Like the Japanese keiretsu, the chaebol tend to provide in-house finance, construction companies to build new plants, transport companies, engineering to supply production equipment, etc. Thus, the Daewoo group incorporates Chungbuk Bank, for example. Kia is the only vehicle producer to fall outside the chaebol system. However, it is part of a small grouping that comprises some key industrial activities. These closely integrated industrial combines give great strength to the vehicle activities. The enormous power (not least financial) of Daewoo as a chaebol, for example, allows the ambitious entry strategies of Daewoo Motors into Europe, with the company setting up much of its own network in Germany and the UK. Although a Japanese keiretsu has similar benefits, it does not have the in-house expertise and ability to spread its risk over many diversified sectors.

The chaebol structure is therefore interesting, because it is one which allows growth into new sectors of activities and offers to such large groups the possibility of synergies between the various activities — synergies we believe will be important in the turbulent technological future the motor industry is facing.

The limits of globalisation?

Globalisation has become a core strategic concept for the automotive industry as it tries to recapture the economies of scale needed to run the Buddist paradigm. Yet several of the trends that we have identified in this book show countervailing forces that will limit the extent to which 'world car' concepts and global-scale integration will constitute a viable long-term strategy. We showed in Chapter 2 that markets in the three major economic regions were not homogeneous or directly comparable, with significant differences in the structure of those markets and the characteristics of the cars sold. We further alluded to the rather different requirements that 'motorisation' may express in other markets in emerging economies around the world.

In Chapter 3 we showed how regulation and government-funded collaborative R&D initiatives were also quite distinct in the three major regions, and likely to yield very different outcomes. Chapter 4 showed that, among other features, alternative materials resources were not well developed in Japan, and that North America was particularly well endowed with aluminium supplies. Elsewhere throughout the book we have discussed the uneven way in which technologies have spread through the world car industry. In short, a single, globalised and uniform industry does not look likely.

On the contrary, the developments that we have outlined imply a fragmentation of the industry and perhaps the emergence of new players in the electric vehicle sector.

9.5   THE PLUTO EFFECT AND TRANSITION SOLUTIONS

Radical change is still not guaranteed. Mom & van der Vinne (1995: 45) refer to the Pluto effect. The term is derived from Mickey Mouse's pet dog; with a sausage dangling in front of him, he will keep following the sausage without ever reaching it. Technological change is similar, the authors argue. A threatened technology, such as the existing car paradigm, will try and adopt as many characteristics of the most likely rival alternative technologies as possible in order to negate the advantages of the latter. This leads to a range of transition solutions.

An example is the EV technology outlined in Chapter 5. We saw there that battery weight is a major problem. This, and the need to reduce fuel consumption, has led to a range of experiments and practical applications of alternative materials solutions to reduce body weight. These alternative materials have become the new technology. In turn, steel then shows that it too can adopt a weight-reduction strategy. However, ULSAB shows that in doing this it adopts many of the characteristics in terms of body construction of the alternative technology. This can entail a move towards a more spaceframe-like approach: Fiat's Multipla, for example, uses a steel spaceframe with steel cladding.

The net result of the Pluto effect is that the differences between the old and the new technology become so small that the advantages of the new paradigm become too small to justify the paradigm shift to the new technology, thus saving the old paradigm. In other words, the new technological paradigm is constantly being pursued without ever being reached.

9.6   THE CHANGING NATURE OF AUTOMOBILITY IN THE 21ST CENTURY

The environmental pressures on the industry are likely to increase over the next decade and, beyond the direct legislative pressures, ordinary citizens are likely to have greater environmental awareness and concerns. As Glaister (1996: 12) puts it: 'Ugly conflict is brewing as the citizen as motorist has increasingly begun to clash with the citizen as inhabitant, as each makes conflicting demands for "better quality of life".' He argues further (Glaister, 1996: 12) that more and more people are questioning 'whether the environmental and social costs of building roads — air pollution, noise and congestion in towns — can be justified'.

In the UK we have already witnessed increasing resistance from direct action fringe groups, such as the Dongas 'tribe' and 'Reclaim the Streets' (see Penman, 1995), enjoying increased media and public support in the absence of clear, decisive government action. The conflict is also one between individualism and society.

The car has been a major individualising force (Freund & Martin, 1993). However, this development really started with the bicycle, as Herrestal (1994) points out. For the first time the citizen had access to a means of transport that was truly independent in the way that even a horse was not (Figure 9.2). It was the liberation of this kind of individualism that allowed motoring to develop. Even today, the bicycle provides greater freedom and independence than the car, in that it is still less regulated and allows access to areas no car can reach, particularly since the advent of the mountain bike. In addition, it is not

Figure 9.2 With the Bicycle the Citizen had Access for the First Time to a Means of Transport that was Truly Independent

dependent on finite fuel resources in its use. The relatively limited nature of the freedoms the car provides is not always appreciated, but as it begins to dawn on an increasing number of motorists and as those freedoms are — of necessity — further eroded, the myth of the car as liberator may collapse.

The car culture is in fact much more fragile than it may appear at present. As people's attentions and spending power embraced the car, so at least some consumers may abandon the car to embrace other, newer and more exciting technologies. In fact, many young people are more interested in computers than cars. This fragility was nicely summarised in a surprising source by Worland (1994): 'Choice, refinement and status — words that symbolise the essence of every essentially unnecessary consumer product, products that signify further progress towards the Utopian goal that forces us all to spend to the maximum of our disposable incomes.'

A combination of declining disposable incomes and a range of new technologies in other areas from which to choose may gradually marginalise the car and reduce it to 'white goods' status, unless cars begin to embody new technologies and recapture that spirit of excitement by becoming more in tune with the new thinking, social changes and dynamic developments in technology. At the same time, the in-use phase is becoming more important in the light of environmental concerns. The fact that most European car buyers buy used rather than new further emphasises the misguidedness of the industry's focus on new car sales. In France, sales of used cars outstripped those of new cars by a factor of

2.14 in 1995 (L'Argus, 1996: 100); a fairly typical ratio for mature European markets. In the UK in 1994, the value of the used-car market exceeded that of the new-car market for the first time and, in 1995, continued to grow ahead of the new-car market, emphasising this trend (ADT, 1996). All these developments require a radical change for a century-old industry.

For the first time also, environmental pressure groups are getting directly involved in the car design and development process. Greenpeace showed a heavily modified version of the Renault Twingo, named Smile, in the summer of 1996. The car had a new smaller supercharged engine with other improvements in aerodynamics and weight reduction, giving an improvement in fuel consumption over the existing Twingo of some 43% (Schoon, 1996). The vehicle is capable of an average fuel consumption of 3.2 litres per 100 km, it uses no radically different technologies and could easily be built today for a premium — according to Greenpeace — of only 12%, or US$1350 (Lewin, 1996).

Changes in attitudes are already beginning, with some marginally committed motorists opting to abandon the car altogether. Some swap their cars for bikes, while others rely on public transport and perhaps occasionally renting a car. An intermediate option is car-sharing schemes, already widespread in Germany (cf. Loste, 1996, for the Berlin scheme), Switzerland, parts of Austria and some Dutch cities, and now also beginning to appear in the UK. One of the first of such schemes in the UK is starting in Edinburgh (Riddoch, 1996). This scheme is modelled on the one in Bremen where 800 people share 48 cars for a monthly fee of around £10 plus £1.40 per hour and 17p per kilometre (Freeman, 1996; Mega & Ramesh, 1996). As Meijkamp & Theunissen (1996: 2) point out, a spread of car sharing will reduce overall car demand. On the other hand, it will enhance the role of the aftermarket, especially dealers and car-rental companies. Change in attitudes and behaviour — as in the Indonesian case outlined above — is crucial, but change would also be required in our land-use practices, with more compact cities (Bevins, 1996) and a halt to out-of-town and edge-of-town shopping developments. The latter can have a measurable impact on local pollution and congestion levels, as was shown with the Gyle shopping centre to the west of Edinburgh (McCann, 1996).

Similarly, after examples in the US (cf. Nieuwenhuis et al., 1992: 116), experiments with car pooling have started all over Europe. Apart from reducing congestion, reducing journey times and wear on the individual cars, thus boosting residual values, this can have a socialising effect, thus turning back some of the adverse social effects of the car.

At the same time, much of the developed world is entering a phase of a rapidly ageing population. By 2021, more than half the European adult population will be over 50 (Kingston, 1996). This may mean a population more attuned to motorisation and less mobile without it. It also has direct implications for the types of cars the market demands. Older people tend to favour smaller cars (Van Leeuwen, 1996: 29), although they are not necessarily less receptive to radical innovation. Renault has been surprised at the old age profile of buyers of the

Twingo. This thinking has also entered the realms of car design and the Royal College of Art has been paying particular attention in recent years to this 'neglected but increasingly powerful consumer group', and has organised meetings with students and older people where the students have been surprised at 'the older group's eagerness to exploit new technologies' (Kingston, 1996).

The downsizing trend is already beginning in Europe. The EIU (Genet, 1996b) forecasts that consumers will increasingly be switching to smaller cars, partly as a result of greater choice, especially in the lower segments. Car makers have been neglecting this segment over the past few decades, leading to an ever smaller number of innovative yet aged designs such as the Mini and 2CV. This new dynamism, fuelled by cars such as the Fiat Cinquecento, Renault Twingo, VW Polo, Ford Ka (Figure 9.3) and others will lead, according to the report, to increasing competition and even more pressure on car makers' profitability. The celebrated car designer Giorgetto Giugiaro (1996) has proposed setting a maximum limit on a car's length of 4.5 or 4.6 m, although most European cars already fall within this limit.

In Chapter 1 we mentioned the anti-car measures increasingly slowing down growth of the car parcs of countries in the Far East. Such measures may also appear in Europe before long. In fact, the more densely populated markets in Europe show signs of early saturation similar to those of Far Eastern markets such as Japan. According to automotive forecasters DRI/McGraw-Hill in London, who tend to base their automotive forecasts primarily on macro-

Figure 9.3 The Ford Ka is One of a New Generation of Small Cars Launched by European Car Producers in the 1990s

economic data, the Dutch market should be able to absorb some 650 000 cars a year, rather than the by now fairly typical 446 388 that were sold in 1995 (Van Leeuwen, 1996: 29). As in Japan, the reasons for this are clear and relate to high population density and comprehensive public transport infrastructure, as well as intensive bicycle use for shorter trips.

9.7   THE INTELLIGENT HIGHWAY: THE DEATH OF MOTORING?

The so-called 'intelligent highway' (or Intelligent Vehicle Highway System — IVHS) is often put forward by the industry as the solution to the traffic congestion problem, and presents a politically attractive option in that it entails intensification of the use of existing capacity without new road construction. By optimising the available road capacity, mass motorisation may remain viable for a bit longer. However, there are also severe dangers associated with this approach. Whitelegg (1993: 144), for example, is highly critical: 'In the case of in-car navigation we have nothing more than pre-programmed, commercial, rat-running.'

Indeed, in the case of UK motorways, if there is a problem on a particular stretch of motorway, it is difficult to see alternative routes to divert on to that do not increase the burden of traffic on urban or rural areas ill-prepared for the increase in traffic. Unlike Tokyo, for example, there are few situations in the UK where traffic could be guided onto a parallel stretch of motorway.

Another problem is that there are suggestions coming out of the Prometheus programme that the infrastructure could control the vehicle, thus relieving the driver. Regulators are attracted by the contribution such technologies could make to road safety, as well as the enhancement of overall car transport efficiency through interlinked vehicles forming a sort of 'road train'. We saw in Chapter 6 that the SMMT/Coventry University Concept 2096 is based on this premise. This would turn the car into a component of a public transport system with high-speed trains of electronically guided and connected vehicles travelling our motorways. This would truly mean the end of motoring as it is popularly understood. It would finally kill off the illusion of freedom associated with the motor car from its inception. It would turn the car from a private mode into a public mode with occasional, if any, private use. The final refuge for the individual would be the private bicycle or less regulated human-powered vehicle.

Many technical developments are being pursued, with the Japanese in particular showing a considerable interest in the application of advanced sensors and electronic control systems. Typical approaches include systems which automatically steer a vehicle between two white lines (underground magnetic cables have also been suggested); systems which use infra-red or radar to keep a safe distance from the car ahead; and sensors which detect another vehicle in close proximity and alert the driver. While in most cases the driver is allowed to override the system, the whole thrust of these developments is to reduce the input from the driver.

Fortunately, in the shorter term, car manufacturers are reluctant to pursue this route because of the potential liability problem. If a car crashes while it is under the control of the vehicle guidance system, who is liable? Clearly not the driver, as he was not in control. But is it the infrastructure provider or the car manufacturer? In this context it is perhaps surprising how supportive car makers have been of such systems, considering their clear potential for centralised control and hence rendering obsolete the basic principles of the private car.

There are clearly limitations to the IVHS and 'intelligent safety' systems in the car. Most of the automatic vehicle guidance systems only work in the relatively simple environment of a motorway. Detailed in-vehicle route guidance systems are only of limited value, although this market is growing. Interactive traffic management systems can only work within the constraints imposed by the existing vehicle infrastructure and the numbers of vehicles seeking to use it.

Motorists will be increasingly excluded from the cities. They will face ever stricter control and surveillance regimes on inter-urban motorways. For many in the advanced industrial world, the myth of the freedom of the open road has long since been exposed as a fanciful — if sometimes attainable — piece of advertising. As the myth evaporated, the car changed in status from indicator of material wealth to fashion accessory. With the advent of IVHS and other measures to restrict individual freedom of movement in cars, we shall truly see the death of motoring.

9.8 CONCLUSIONS

The car may be described as the most complete expression of 20th century culture. However, how appropriate will it be for the 21st century? Although car ownership in itself is still a status symbol in the emerging economies such as South Korea, in the West car ownership has come to be taken for granted. We are seeing the end of a phase where the type of car, its 'designer label', determines status and the beginning of a phase where other criteria play a more important role. The central role of the car in people's lives is becoming less passionate. The magic of the car is disappearing as the younger generation hanker less after car ownership or their driving licence than did their parents. The car as white good, as a means of getting from A to B, is around the corner; is there really a need for a new Model T (Van Leeuwen, 1996: 32)? The car as white good, however, may be preceded or accompanied by the car as fun object. The downsizing trend towards the Twingo, MCC Smart car and even Mercedes A-class illustrate this.

Unfortunately, any revolution has its victims and, in this case, this will be in the form of jobs. As the old industry restructures, jobs are likely to be lost; a trend that has already been ongoing for some time, as we saw earlier in this book. This pressure on the existing system is also illustrated by the following statement from PSA chairman Calvet: 'We will not close any major plant. Nevertheless, there will be problems with the size of our workforce, given our productivity requirements' (Johnson & Farhi, 1996).

For the people concerned, a critical issue of course will be the geography of growth and decline. The new technologies may well be more labour-intensive and thus able to absorb some people in the transition period. On the other hand, a key element of the transition to sustainable motorisation will surely be much lower volume production and considerably extended product longevity. Both technical developments — in steel as well as the increasing popularity of alternative materials — and the need to reduce the waste stream will force the durability issues to the top of the agenda within the next decade (Deutsch, 1994). Electric vehicles also tend to have a longer lifespan. Although batteries have a more limited service life, electric motors last a long time. Recycling is not the answer, as it generally leads to downgrading and does not stop waste. It also requires energy and produces pollution, as we saw in Chapter 7.

In aggregate, employment then is bound to fall. Moreover, EOVs would magnify the existing trend toward greater product reliability and lower service needs — with the result that the support infrastructure required would be smaller than that which currently exists. Again, employment is bound to fall. On the other hand, the need for retrofitting, generally dealing with a much older parc, managing body modules, running leasing programmes and other aftermarket developments are likely to absorb a certain number of people.

The changes we set out in this book have already begun. Many of the trends can be recognised and are unstoppable. We have the choice between continuing our present course towards increasingly heavy and over-specified cars, ultimately controlled by intelligent highway systems — the death of motoring — or a move towards lightweight long-life EOVs, designed as true driving machines, but leading to the death of the conventional car industry. The impact on the industry could be devastating, but may generate at the same time a new dynamism and period of creativity as the industry gradually adapts to a more sustainable format. However, we saw with the spread of mass production and the Buddist paradigm that such changes can take a few decades to come about. It may only be with hindsight that we will realise in 30 years' time or so just how dramatic the change has been.

References

AAMA (1995) *World Motor Vehicle Data*. Washington, DC: American Automobile Manufacturers Association.

AAMA (1996) *World Motor Vehicle Data*. Washington, DC: American Automobile Manufacturers Association.

Abe, K, A Hayashi & M Sakaino (1995) An update on Nissan Intelligent Body Assembly System, International Body Engineering Conference 1995. *Body Assembly and Manufacturing*, **15**, 1–7.

ACEA (1994a) Car recycling: finding a strategy for the treatment of end-of-life vehicles. *ACEA Newsletter*, **11**, 1–9. Brussels: Association des Constructeurs Européens d'Automobiles.

ACEA (1994b) Co-operative automotive research and development: the route to optimisation. *ACEA Newsletter*, **16**, 1–5.

ACEA (1994c) Soothing words from Korea. *ACEA Newsletter*, **15**, 1–5. Brussels: Association des Constructeurs Européens d'Automobiles.

ADT (1996) *The Used Car Market*. London: ADT Auctions and Paragon Communications.

AEA (1995) *The Aluminium Car*. London: Aluminium Extruders Association.

Aglietta, M (1979) *A Theory of Capitalist Regulation: The US Experience*. London: New Left Books.

AID (1995) Saturation in Sight, *Automotive Industry Data Newsletter*, 9519, 1–2.

AISI (1994) *Holistic Design with Steel for Vehicle Weight Reduction*. Michigan: American Iron and Steel Institute.

Aluminum Association (1995) Aluminium structured vehicles. *Auto Design Review*, special issue.

Anon. (1992) Progress report: recent developments in new steel technologies. *Metal Producing*, **10**, 25–34.

Anon. (1996a) *Automotive News Data Book*. Detroit: Automotive News.

Anon. (1996b) Options breakdown. *Autocar*, 31 July, 14.

Baldensperger, D (1996) La France industrielle: une histoire dans laquelle l'automobile a voix au chapitre. Book review, *L'Argus de l'Automobile*, 24 October, 31.

Bannister, D & K Button (eds) (1993) *Transport, the Environment and Sustainable Development*. London: Spon.

Baysore, J (1995) Quality and performance drive laser welded blank applications. *Proceedings IBEC '95, Materials and Body Testing*, 116–19.

Bellu, S (1989) Il concetto dei 'mutanti'/The 'mutant' concept. *Auto & Design*, July, 42–9.

Berger, R & H-G Servatius (1994) *Die Zukunft des Autos hat erst Begonnen; Oekologisches Umsteuern als Chance*. Munich: Piper.

Bevins, A (1996) A Coronation Street for every town. *The Independent*, 22 August, 3.

Bhaskar, K (1979) *The Future of the UK Motor Industry*. London: Kogan Page.

Bonnaud, C (1995) Design: l'image virtuelle au pouvoir. *L'Auto-Journal*, **404**, 16 February, 40–1.

Bonnett, R, R Carpenter, D Loosle & R Provancher (1995) Structural composites in a lightweight body structure. *Proceedings IBEC '95, Body Design and Engineering*, **13**, 38–44.

Bott, H, H Braess, W Assmann, H Burst, K Federspiel, W Finkenauer, R Georgi, M Grahle, D Hard, L Hamm, W Heck, U Hickmann, M Kessler, R List, H Rombold, B V

Rotberg, K-M Stehle, E Strehler, R Weber, K Bellmann, I Schött, W Schunter, K Schmitt & R Willeke (1976) *Long-Life Car Research Project, Final Report Phase I: Summary.* Stuttgart: Dr Ing hc F Porsche AG.

Brambilla, C-O (1990) La forma 'elettrica'/'Electric shape'. *Auto & Design*, April–May, 71–6.

Brown, S, S Steinstra & K Ludwigsen (1994) *Closing the Loop — The Car Recycling Challenge.* London: Euromotor Reports.

Calliano, A (1995) L'habitat ecologico/Ecological habitat. *Auto & Design*, June–July, 55–61.

Carson, R (1962) *Silent Spring.* London: Hamilton.

CCFA (1994) *L'Automobile à l'aube du XXIeme siècle.* Paris: Comité des Constructeurs Français d'Automobiles.

CD & T (1991a) Audi Avus Quattro. *Car Design & Technology*, December, 18.

CD & T (1991b) Aluminium Audi. *Car Design & Technology*, November, 20–1.

CD & T (1991c) Volkswagen Chico. *Car Design & Technology*, November, 24–5.

CEC (1992) *The Future Development of the Common Transport Policy.* Brussels: Commission of the European Communities, COM(92)494 Final.

CEC (1996) *A Community Strategy to Reduce CO_2 Emissions from Passenger Cars and Improve Fuel Economy* (Communication from the Commission to the Council and the European Parliament). Discussion document, Brussels.

Chapman, G (1996) Happy birthday to the car. *Earth Matters*, **29**, Spring, 8–11.

Chatterjee, A (1995) Recent developments in iron and steelmaking. *Iron and Steelmaking*, **22**(2), 100–4.

Chrysler (1996) At http://www.hev.doe.gov/program/ch_sub.html

Clegg, A (1996) Promoting sustainable manufacture through Environmental Management System Standards. Paper presented at the 5th International Greening of Industry Network Conference, 'Global Restructuring: A Place for Ecology?', Heidelberg, 24–27 November.

Cooper, T (1994) *Beyond Recycling: The Longer Life Option.* London: The New Economics Foundation.

Coöperatieve Vereniging Witkar (1974) *Technical Information Regarding the Proposed Demonstration Project.* Amsterdam: CVW.

Coöperatieve Vereniging Witkar (1975) *Witkar Nu.* Amsterdam: CVW.

Coventry Electric Vehicle Project (1996) *The Coventry Electric Vehicle Project.* Coventry: Peugeot.

Cronk, S (1995) *Building the E-Motive Industry: Essays and Conversations About Strategies for Creating an Electric Vehicle Industry.* Warrendale, PA: Society of Automotive Engineers.

Courteauld, P (1991) *Automobiles Voisin 1991–1958.* London: White Mouse Editions & Saint Cloud, EPA.

Das, R & R Das (1995) *Wegen Naar de Toekomst.* Baarn: Tirion.

Davies, P (1994) *Road Vehicles: Report to the Management Committee on the Case for a Research Programme in the Manufacturing Sector of Road Vehicles.* London: The Innovative Manufacturing Initiative.

Delerm, J-C (1986) L'automobile, une des plus grandes aventures humaines. In *Le Musée National de l'Automobile*, Mulhouse.

DeSimone, L D (1996) On http://www.mmm.com/profile/pressbox/contract.html

De Telegraaf (1996) Hybride te koop. *De Telegraaf*, 25 October, 22.

Deutsch, C (1994) *Abschied vom Wegwerfprinzip: Die Wende zur Langlebigkeit in der industriellen Produktion.* Stuttgart: Schaeffer-Poeschel.

DGXVII (1996) European energy to 2020: a scenario approach. In *Energy in Europe* (special edition). Directorate General for Energy, Luxembourg: Office for Official

Publications of the European Communities.

Dieffenbach, J, A Mascarin & M Fisher (1993) Cost simulation of the automobile recycling infrastructure: the impact of plastics recovery. *SAE Technical Paper 930557*. Warrendale, PA: Society of Automotive Engineers.

Diekmann, A (1994) The debate on the external costs of the automobile: on the wrong track. *ACEA Newsletter*, 15, 6–11.

Diem, W R (1992) Cost is biggest question, most elusive answer. *Automotive News*, 12 October, 34.

Dillen, A (1994) Europese overheid moet auto-industrie steunen. *Auto Nieuws*, 9 March, 7.

EAE (1995) *Environment in the European Union*. European Environment Agency, Luxembourg: Office for Official Publications of the European Communities.

Ealey, L & T Gentile (1995) The potential impact of electric vehicles on the automotive business system. *EIU International Motor Business*, 2nd quarter. The Economist Intelligence Unit, 102–15.

Eckerman, E (1989) *Automobile: Technikgeschichte im Deutschen Museum*. Munich: Beck.

Elkington, J & J Hailes (1988) *The Green Consumer Guide*. London: Gollancz.

Elton, B (1991) *Gridlock*. London: Sphere Books.

European Commission (1996) *A Community Strategy to Reduce CO_2 Emissions from Passenger Cars and Improve Fuel Economy*. Communication from the Commission to the Council and the European Parliament.

Fabre, F, A Klose & G Somer (eds) (1987) *COST 302 Technical and Economic Considerations for the Use of Electric Road Vehicles*. Luxembourg: DGVII and DGXII, CEC.

Farhi, S (1996) France asked for job cuts. *Automotive News Europe*, 22, 1.

Farrington, D & H Ryder (1993) The environmental assessment of transport infrastructure and policy. *Journal of Transport Geography*, 1(2), 102–18.

Feast, R (1996) If anyone can lose money, Nissan can. *Motor Industry Management*, September, 4.

Fiedler, J (1992) *Stop and Go: Wege aus dem Verkehrschaos*. Cologne: Kiepenheuer & Witsch.

Fleet News (1994) Used car market hits 18bn record. *Fleet News*, 29 July, 6.

Florida, R & M Kenny (1993) The new age of capitalism. *Futures*, 25(6), 636–51.

Ford (1996) At http://www.hev.doe.gov/program/ford_sub.html

Ford & Alcan (1994) *Aluminum Intensive Vehicle*. Detroit: Ford.

Freeman, C (1993) Technical change and future trends in the world economy. *Futures*, 25(6), 621–35.

Freeman, V (1996) Instant rentals on your doorstep. *CAR 96*, 10 February.

Frere, P (1992) Continental diary. Turin: the Cinquecen Toy show. *Car Design & Technology*, July, 16–17.

Freund, P & G Martin (1993) *The Ecology of the Automobile*. Montreal: Black Rose Books.

Fukukawa, S (1992) Japan's policy for sustainable development. *Columbia Journal of World Business*, XXVII (iii, iv), 96–105.

Gadacz, O (1996) Samsung will spend $13 billion to build cars. *Automotive News Europe*, 19 February, 15.

Genet, J-P (1996a) Renault: des autos non rentables et pourtant encore trop cheres. *L'Argus de l'Automobile*, 28 March, 32.

Genet, J-P (1996b) En Europe, l'avenir est aux petites voitures, *L'Argus de l'Automobile*, 7 Sept., 9.

Georgano, G (ed.) (1968) *The Complete Encyclopedia of Motorcars, 1885–1968*. London: Ebury Press.

Giannini, M (1983) Volvo guarda al duemila/Volvo, heading towards the year 2000. *Auto*

& *Design*, December–January 1984, 60–4.

Giugiaro, G (1996) Giugiaro suggests real downsizing of cars. *Automotive News*, 8 January, 14.

Glaister, S (1996) How to kick the car. *Design*, Autumn, The Design Council, 12–14.

GM (1996) General Motors at http://www.hev.doe.gov/program/gm_sub.html

Gormezano, J, J Hartley, B Knibb, I Ruxton & P Nieuwenhuis (1992) *The Auto Industry: Towards a Cleaner Environment*. Derby: Knibb, Gormezano & Partners.

Gouldsen, A (1996) Environmental policy and industrial competitiveness: searching for synergy. Paper presented at the 5th International Greening of Industry Network Conference, 'Global Restructuring: A Place for Ecology?', Heidelberg, 24–27 November.

Government of Japan (1996) *Comprehensive Plan for ITS in Japan*. Tokyo: Government of Japan.

Grayson, S (1978) The all-steel world of Edward Budd. *Automobile Quarterly*, XVI(4), Fourth Quarter.

Hamel, G (1991) Competition for competence and inter-partner learning within international strategic alliances. *Strategic Management Journal*, 12, 83–103.

Hamm, L (1993) Steel — a promising option in future vehicle body construction? Paper presented to 'IISI 27: Advanced Steel Design: Downweighting the Automobile', Paris.

Hanicke, L (1995) Ten years of experiments with high power lasers in production. *Proceedings IBEC '95, Body Assembly and Manufacturing*, IBEC, Warren, 66–75.

Have, R T (1993) How Amsterdam limits car use. Paper presented to conference, 'The Automobile Industry and Society at the Crossroads', Amsterdam School of International Relations, 7 May.

Heller, R (1996) Kings of the road. *Design*, Autumn, The Design Council, 4–5.

Herrestal, A (1994) The bicycle and the rise of individualism. *Bike Culture Quarterly*, 3, August, 170–3.

Hillman, H, J Adams & J Whitelegg (1990) *One False Move: A Study of Children's Independent Mobility*. London: Policy Studies Institute.

Hoeven, J, F Lambert, K Rubben, I De Rycke & E Leirman (1995) Design and manufacturing issues for components made from tailored blanks. *Materials and Body Testing*, International Body Engineering Conference, IBEC, Warren, 96–107.

Hoogma, R, R Kemp, G Praetorius & D Truffer (1996) Organising for technological regime shifts in road transport: societal experiments in regional contexts. Workshop D4, 5th International Research Conference of the Greening of Industry Network, Heidelberg, 24–27 November.

Hughes, R (1995) Lightweight automotive design: the ultralight steel auto body. *Proceedings IBEC '95, Body Design and Engineering*, 59–73.

IISI (1994) *Competition between Steel and Aluminium for the Passenger Car*. Brussels: International Iron and Steel Institute.

JAMA (1995) *The Motor Industry in Japan*. Tokyo: Japan Automobile Manufacturers Association.

Janicki, G (1992) *Cars Europe Never Built*. New York: Sterling.

Jatoorapreuk, B (1996) Decentralisation and introducing mass transit. *The Wheel Extended*, 96, 14–15.

Johnson, R & S Farhi (1996) Calvet: partial alliances will grow in Europe. *Automotive News Europe*, 9 December, 11.

JTERC (1995) *Transport Policy Prospects for the 21st Century*. Tokyo: Japan Transport Economics Research Centre.

Keebler, J (1995) Chrysler tests standardized assembly tools. *Automotive News*, 20 March, 24B.

KERI (1994) *Korean Automotive Industry, 1994*. Seoul: Kia Economic Research Institute.

Kewley, D, I Cambell & J Wheatley (1987) Manufacturing feasibility of adhesively bonded aluminium for volume car production. *SAE Technical Paper 870150*. Warrendale, PA: SAE.

Kingston, P (1996) Pole position at seventy plus. *Guardian Education*, 1 October, 2.

Kleese, E, D McBain & E Haque (1995) Listening to the customer's voice: the development of GenCorp's flexible SMC system (Flexion). *Proceedings IBEC '95, Materials and Body Testing*, 14, 15–19.

Kubota, H (1996a) World's worst traffic jams and their causes. *The Wheel Extended*, 96, 5–9.

Kubota, H (1996b) The world's most congested city, and how local residents move around. *The Wheel Extended*, 96, 10–13.

Kurylko, D (1996) VW turnaround shows up in profits. *Automotive News Europe*, 11 November, 3.

L'Argus (1995) *Annual Statistical Review*. Paris: L'Argus de l'Automobile et des Locomotions.

L'Argus (1996) *Numero Special 96*. Paris: L'Argus de l'Automobile et des Locomotions.

Lamming, R (1994) *Beyond Partnership*. London and New York: Prentice Hall.

La Vie Automobile (1946) Le salon de 1946. *La Vie Automobile*, 25 September, 129–39.

Lee, K-H (1994) *Change Begins With Me — Samsung's New Management*. Seoul: Office of the Executive Staff of the Samsung Group.

Lee-Harwood, B (1996) Cars of the future. *Earth Matters*, 29 (Spring), 12.

'Lei' (1986) Neuer Anlauf. *Auto Motor und Sport*, 6, March.

Leniger, H (1996) Frankrijk: Nieuwe ontwikkelingen voor de TGV. *Technieuws*, 34(4), 23–6.

Lewin, T (1996) Greenpeace car uses new engine. *Automotive News Europe*, 10 June, 23.

Lindh, B-E (1984) *Volvo: The Cars — From the 20s to the 80s*. Malmo: Forlagshuset Norden.

Loste, F (1996) StattAuto, a Berlin, met des voitures en copropriété. *L'Argus de l'Automobile*, 5 September, 12.

Lovelock, J (1979) *Gaia: A New Look at Life on Earth*. Oxford: Oxford University Press.

Lovins, A (1995) Hypercars: The Next Industrial Revolution. Keynote address to the 1995 Automobile Distribution and Servicing Conference, Brussels, 5 December.

Lovins, A, J Barnett & H Lovins (1993) *Supercars: The Coming Light-Vehicle Revolution* (scholarly preprint in advance of conference proceedings). Snowmass, CO: Rocky Mountain Institute.

Lowe, M (1989) *The Bicycle: Vehicle for a Small Planet*, Worldwatch Paper 90. Washington, DC: Worldwatch Institute.

Made in Fiat (1996) Turin leads Europe in electric car promotion. *Made in Fiat*, October, Fiat UK Ltd.

Margolius, I & J Henry (1990) *Tatra — The Legacy of Hans Ledwinka*. Harrow: SAF.

Marien, M (1992) Environmental problems and sustainable futures. *Futures*, 24(8), 731–57.

Martin, D & P Peterson (1995) Ultralight steel auto body: how 32 competitors found harmony in pursuing a common goal. *Proceedings IBEC '95, Advanced Technologies and Processes*, IBEC, Warren, 18–20.

Mascarin, A, J Dieffenbach, M Brylawski, D Cramer & A Lovins (1995) Costing the Ultralite in volume production: can advanced composite bodies-in-white be affordable? Paper presented at the 1995 International Body Engineering Conference and Exposition, Detroit, 31 October–2 November.

Mathews, A (1994) Role of steel in fuel efficient cars. *Iron and Steelmaking*, 21(2), 87–92.

Mathews, A, D Wheeler & H Jordan (1993) Fuel efficient sheet steel automobiles. *SAE Technical Paper 930785*. Warrendale, PA: SAE.

Maxton, G & J Wormald (1995) *Driving over a Cliff? Business Lessons from the*

World's Car Industry. Wokingham: The Economist Intelligence Unit & Addison-Wesley.

Maxwell, J (1995) Self-regulation and social welfare: the political economy of corporate environmentalism. *Proceedings of the 1995 Business Strategy and the Environment Conference*, University of Leeds, Shipley, ERP Environment, 20–21 September.

McCann, A (1996) Gyle traffic is a major cause of pollution. *The Scotsman*, 17 August, 5.

McGregor, I, D Nardini, Y Gao & D Meadows (1992) The development of a joint design approach for aluminium automotive structures. *SAE Technical Paper 922122.* Warrendale, PA: SAE.

McGurn, J (1996) Top marks to Denmark. *Bike Culture Quarterly*, 10, July, 38–40.

McKinlay, A & K Starkey (1994) After Henry: continuity and change in Ford Motor Company. *Business History*, 28(1), 184–205.

McKinstry, S (1993) Financial management in the Scottish automobile industry. *Accountancy, Business and Financial History*, 3(3), 275–90.

Meana, C (1992) The shape of the EC and sustainable development. *Columbia Journal of World Business*, XXVII (iii, iv), 106–11.

Mega, M & R Ramesh (1996) Council launches car rental club to reduce city centre congestion. *The Times*, 21 January.

Meijkamp, R & R Theunissen (1996) 'Car sharing': consumer acceptance and changes on mobility behaviour. Paper presented at the 24th European Transport Forum, London, 2–6 September.

Metzner, H (1995) Global warming: stop rushing to judgement. *ACEA Newsletter*, 22, 8–10.

MIRU (1990) Into the 1990s: *Future strategies of the vehicle producers of South Korea & Malaysia*, Norwich, Motor Industry Research Unit.

Moerman, P & B De Bleser (1995) Long time production experience with high power CO_2 lasers in body shop. *Proceedings IBEC '95, Body Assembly and Manufacturing*, IBEC, Warren, 57–65.

Mom, G & V van der Vinne (1995) *De Elektro-auto: een Paard van Troje?* (The electric car: a Trojan horse?). Deventer: Kluwer.

Nadis, S & J MacKenzie (1993) *Car Trouble*. Boston, MA: Beacon Press.

Nardini, D & A Seeds (1989) Structural design considerations for bonded aluminium structured vehicles. *SAE Technical Paper 890716*. Warrendale, PA: SAE.

Nelson, D & R Metzger (1995) Advanced transfer feed technology. *Proceedings IBEC '95, Body Assembly and Manufacturing*, IBEC, Warren, 15, 90–7.

Newton, S & D Iddiols (1993) From hearses to horses: launching the Volvo 850. *Journal of the Market Research Society*, 145–61.

Nieuwenhuis, P (1992) The operating regime and infrastructure for electric vehicles. Paper presented to IMechE Seminar Battery Electric and Hybrid Vehicles. London: Institution of Mechanical Engineers, 10–11 December.

Nieuwenhuis, P (1994a) The environmental implications of Just-In-Time supply in Japan — lessons for Europe? *Logistic Focus*, 2(3), 2–4.

Nieuwenhuis, P (1994b) Emissions legislation and incentives in the USA and Europe. In P Nieuwenhuis & P Wells (eds), *Motor Vehicles in the Environment: Principles and Practice*. Chichester: John Wiley & Sons.

Nieuwenhuis, P (1994c) The long-life car: investigating a motor industry heresy. In P Nieuwenhuis & P Wells (eds), *Motor Vehicles in the Environment: Principles and Practice*. Chichester: John Wiley & Sons.

Nieuwenhuis, P (1995) The impact of environmental legislation on the automotive industry. *Proceedings of the 1995 Business Strategy and the Environment Conference*, University of Leeds, Shipley, ERP Environment, 20–21 September, 177–82.

Nieuwenhuis, P & P Wells (1993) A new automotive paradigm for the 21st century:

towards the light-weight, long-life vehicle. Paper presented at Challenging the Environmental & Safety Needs in the Design of Cars, seminar organised by the Institution of Mechanical Engineers, London, 3 December.

Nieuwenhuis, P & P Wells (eds) (1994) *Motor Vehicles in the Environment*. London: John Wiley & Sons.

Nieuwenhuis, P & P Wells (1996) Scrappage incentives: an environmental analysis. *FT Automotive Environment Analyst*, **20** (September), 15–17.

Nieuwenhuis, P, P Cope & J Armstrong (1992) *The Green Car Guide*. London: Green Print.

Norcia, B Cotton & J Dodge (1994) Environmental performance and competitive advantage in Canada's paper industry, *Business Strategy and the Environment* **2** (4), 1–9.

OECD (1992) *Market and Government Failures in Environmental Management: The Case of Transport*. Paris: OECD.

Ohmae, K (1985) *The Rise of Triad Power*. New York: Harper & Row.

Olszewski, P & D J Turner (1993) New methods of controlling vehicle ownership and usage in Singapore. *Transportation*, **20**(4), 355–72.

OTA (1995) *Advanced Automotive Technology: Visions of a Super-Efficient Family Car*. US Congress, Office of Technology Assessment, OTA-ETI-638. Washington, DC: US Government Printing Office.

Penman, D (1995) 'Roads protesters "reclaim" high street' and 'Eco-alliance drives home the anti-car message', *The Independent*, 15 May, 3.

Pirie, M (1990) *Green Machines*. London: Adam Smith Institute.

PNGV (1994) *Inventions Needed for the PNGV*. Washington, DC: Partnership for a New Generation of Vehicles.

PNGV (1995) *Sharing Technology for a Stronger Future*. Washington, DC: Partnership for a New Generation of Vehicles.

Poduska, R, R Forbes & M Bober (1992) The challenge of sustainable development. *Columbia Journal of World Business*, **XXVII** (iii, iv), 286–91.

Polo, J-F (1992) Les villes françaises veulent toutes la voiture électrique. *Les Echos*, 30 September.

Prahalad, C K & G Hamel (1990) The core competence and the corporation. *Harvard Business Review*, May–June, 71–91.

Prothero, A (1994) Green marketing in the car industry. In P Nieuwenhuis & P Wells (eds), *Motor Vehicles in the Environment*. Chichester: John Wiley.

PSA (1992) *Electric Vehicles: Les Cahiers de la Direction de la Communication de PSA Peugeot Citroën No. 1*. Paris: PSA Direction de la Communication.

Pugliese, T (1996) *Global Vehicle Production Trends*, EIU Special Report R336. London: Economist Intelligence Unit.

Raff, D (1991) Making cars and making money in the interwar automobile industry: economies of scale and scope and the manufacturing behind the marketing. *Business History Review*, Winter, 721–53.

Raff, D (1994) Models of the evolution of production systems and the diffusion of mass production methods in the American motor vehicles industry. Paper presented at the 2nd International Meeting of GERPISA, Paris, June.

Rao, B & P Varaiya (1994) Roadside intelligence for flow control in an intelligent vehicle and highway system. *Transportation Research*, **2C**(1), 49–72.

Rawlinson, M & P Wells (1996) Taylorism, lean production and the automotive industry, *Asia Pacific Business Review*, **2** (4), 189–204.

Rechtin, M (1996) Leasing EV1 in US. *Automotive News Europe*, 28 October.

Rees, C (1995) *Three-Wheelers: The Complete History of Trikes, 1885–1995*. Reigate: Blueprint Books.

Renault (1995) *MOSAIC*. Paris: Renault.

Rendell, J (1996) A-class action. *Autocar*, 1 May, 34–7.

Renner, M (1988) *Rethinking the Role of the Automobile*, Worldwatch Paper 84. Washington, DC: Worldwatch Institute.

RERI (1996) *Global Environment and Automobile Traffic*. Tokyo: Road Environment Research Institute.

Rhys, D G (1984) New technology and the economics of the motor industry. Proceedings of International Conference on Future Development in Technology: The Year 2000, Waldorf Hotel, London, 4–6 April.

Rhys, D G, R McNabb & P Nieuwenhuis (1993) The significance of scale in the aftermath of lean production. *EIU International Motor Business*, 1st quarter, The Economist Intelligence Unit, London, 123–50.

Rhys, D G, P Nieuwenhuis & P Wells (1995) *The Future of Car Retailing in Western Europe*. London: The Economist Intelligence Unit.

Riddoch, L (1996) A problem shared . . . . *The Scotsman*, 17 August, 8.

Rigby, R (1996) Planes, trains and automobiles. *Management Today*, September, 104.

Riley, R (1994) *Alternative Cars in the 21st Century: A New Personal Transportation Paradigm*. Warrendale, PA: Society of Automotive Engineers.

Road & Track (1995) *The Complete Road & Track '96 Car Buyer's Guide*. New York: Hachette Filipacchi.

Robert, M (1992) La ville veut fermer ses portes à l'auto. *Les Echos*, 13 December, 15.

Roberts, J, J Cleary, K Hamilton & J Hanna (eds) (1992) *Travel Sickness: The Need for a Sustainable Transport Policy for Britain*. London: Lawrence & Wishart.

Rowlands, I & M Greene (eds) (1992) *Global Environmental Change and International Relations*. Basingstoke: Macmillan.

Royal Commission on Environmental Pollution (1994) *Eighteenth Report: Transport and the Environment*. London: HMSO.

Schampers, B (1994) Overlevingsplan voor Europese auto-industrie. *Eindhovens Dagblad*, 24 February, 6.

Schmarbeck, W (1989) *Tatra: Die Geschichte der Tatra Automobile*. Luebbecke, Uhle & Kleimann.

Schmidheiny, S (1992) The business logic of sustainable development. *Columbia Journal of World Business*, **XXVII** (iii, iv), 18–25.

Schneider, C, W Mueschenborn, G Hartmann & R Bode (1995) Initiatives of steel mills responding to lightweight body construction. *Proceedings IBEC '95, Materials and Body Testing*, IBEC, Warren, 10–14.

Schoon, N (1996) Why is the world's motor industry not willing to produce a car like this? *The Independent*, 14 August, 20.

Schumacher, E F (1973) *Small is Beautiful*. London: Abacus.

Schumpeter, J A (1939) *Business Cycles: A Theoretical, Historical and Statistical Analysis of the Capitalist Process*, two volumes. New York: McGraw-Hill.

Schweitzer, S (1982) *Des Engrenages a la Chaine: Les Usines Citroën, 1915–1935*. Lyon: Presses Universitaires.

Sewells (1995) European model cycles get shorter. *Sewells European Marketing Review*, 8, 10.

Seyad, A, F Senesael & M de Clercq (1996) The use of voluntary agreements as an instrument of environmental policy in the energy sector: a case study of the Belgian Electricity Covenant. Paper presented at the 5th International Greening of Industry Network Conference, 'Global Restructuring: A Place for Ecology?', Heidelberg, 24–27 November.

Seymour J (1996) Trafficking in death, *New Scientist*, 151 (2042), 34–37.

Shacket, S (1981) *The Complete Book of Electric Vehicles*. Northbrook, IL: Domus Books.

SMC Automotive Alliance (1995) *1995 Model Year: Passenger Car and Truck SMC Components*. New York: SMC Automotive Alliance.

SMMT (1995) *New Cars for Old: A UK Vehicle Scrappage Scheme*. London: Society of Manufacturers and Traders.

Stein, R (1962) *The Automobile Book*. London: Paul Hamlyn.

Strassl, H (1984) *Karosserie: Aufgaben, Entwurf, Gestaltung, Konstruktion, Herstellung*. Munich: Deutsches Museum.

Svelander, A (1996) Success for alternative bus fuels in Sweden. *Volvo Globetrotter*, October, 27.

Taragnat, S (1995) Marche des utilitaires: les affaires reprennent. *L'Argus de l'Automobile*, 9 February, 35–6.

The Economist (1996) *Taming the Beast: A Survey on Living with the Car. The Economist*.

Tickell, O (1996) Healing the rift. *Independent on Sunday*, 4 August, 40.

Tims, J (1990) Vehicle emissions. In P Brackley (ed.), *World Guide to Environmental Issues and Organisations*. London: Longman.

Toh, R S (1992) Experimental measures to curb road congestion in Singapore: pricing and quotas. *The Logistics and Transportation Review*, 28(3), 289–317.

Topolansky, A (1993) Alternative fuels. *Columbia Journal of World Business*, XXVII (iv), 38–47.

Topp, H & T Pharoah (1994) Car-free city centres. *Transportation*, 21, 231–47.

Toyota (1992) *Toyota Production System*. Toyota City: Toyota Motor Corporation.

Treece, J (1996a) EVs set to enter main market. *Automotive News Europe*, 28 October, 9.

Treece, J (1996b) EVS-13 hero: RAV4 with a fuel cell. *Automotive News Europe*, 28 October, 9.

Trisoglio, A (1993) International business and sustainable development. In H Bergesen & G Parmann (eds), *Green Globe Yearbook 1993*. Oxford: Oxford University Press, 87–100.

Ujihara, S & D Cooke (1992) The application of laser-textured dull steel to automobile body panels. *Moving Forward with Steel — Automobiles*. London: The Institute of Materials.

ULSAB (1995) *Automotive Steel: A New Approach to Body in White Architecture*. Southfield, MI: Ultra Light Steel Auto Body.

USCAR (1995) *Sharing Technology for a Stronger America*. Detroit: United States Council for Automotive Research.

Van der Burgt, G (1977) Auto's die er nog niet zijn; beschouwingen over het energiethema en het wegverkeer. *THD Nieuws*, Delft, December, 8–17.

Van Dijk, J & J Cramer (1992) *De Auto in Zes Landen*. Amsterdam: Stichting voor Economisch Onderzoek der Universiteit van Amsterdam.

Van Kasteren, J (1977) Zuinig de 21e eeuw in. *THD Nieuws*, December, 11–12.

Van Leeuwen, A (1996) De auto is dood, leve de auto. *Elsevier*, 10 February, 28–31.

Vauxhall (undated) Hybrid Vehicles. *Vauxhall Fact File: Alternative Fuels*. Luton: Public Affairs Dept., Vauxhall Motors Ltd.

Verheul, H & K Termeer (1996) Beyond regulation: government roles in the diffusion of cleaner technologies. Paper presented at the 5th International Greening of Industry Network Conference, 'Global Restructuring: A Place for Ecology?', Heidelberg, 24–27 November.

Volvo (1991a) Environmentally compatible product development with the EPS system, *Environmental Report No. 27*. Gothenburg: Volvo Car Corporation.

Volvo (1991b) *A System for Calculating Environmental Impact*. Stockholm: VCC, Federation of Swedish Industry & Swedish Environmental Research Institute.

VW (undated) *The Volkswagen Environmental Report*. Wolfsburg: Volkswagen AG.

Ward's Autoworld (1995) Special Supplement 'How SMC is shaping tomorrow's vehicles', January.

Wards (1996) *Wards Automotive Yearbook 1996*. Detroit: Wards.

Ware, M (1976) *Making of the Motor Car, 1895–1930*. Ashbourne: Moorland Publishing.

Waters, M (1992) *Road Vehicle Fuel Economy*. Transport and Road Research Laboratory State of the Art Review 3. London: HMSO.

Watson, R, L Gyenes & B Armstrong (1986) *A Refuelling Infrastructure for an All-Electric Car Fleet*. TRRL Research Report RR66. Crowthorne: Department of Transport, Transport and Road Research Laboratory.

Way, A (1996) Gathering depression over Europe. *Motor Industry Management*, September, 2.

WCED (1987) *Our Common Future*, World Commission on Environment and Development. New York: Oxford University Press.

Wells, P. (1996a) The environment and core competence strategy. *Proceedings of the 1996 Business Strategy and the Environment Conference*, Leeds, 19–20 September, 258–63.

Wells, P (1996b) Competitive and collaborative R&D: a comparison of policies for the technological transformation of the automotive industry in the European Union and North America. Paper presented at the 5th Annual Conference of the Greening of Industry Network, 'Global Restructuring: A Place for Ecology?', Heidelberg, 24–27 November.

Wells, P (1996c) The all-steel body: a future beyond the year 2000? *Proceedings of the 4th International Conference Sheet Metal 1996*, University of Twente, 1–3 April, Vol. I, 39–48.

Wells, P (1996d) The carbon tax and the automotive industry in Europe. Paper presented at the 6th Annual European Environment Conference, 'Advances in European Environmental Policy: Critical Issues in Policy Success', Leeds, 16–17 September.

Wells, P & M Rawlinson (1994a) *The New European Automobile Industry*. Basingstoke: Macmillan.

Wells, P & M Rawlinson (1994b) The environment and a traditional sector: the case of the automotive presswork industry. In P Nieuwenhuis & P Wells (eds), *Motor Vehicles in the Environment*. Chichester: John Wiley, 131–51.

Wells, P & P Nieuwenhuis (1996) Fuel economy, CO_2 emissions and the Carbon Tax: the future of motor vehicle regulation in Europe. *Financial Times Automotive Environment Analyst*, **18** (July), 20–2.

Wells, P, P Nieuwenhuis & D G Rhys (1996a) What's happened to the demand for new cars? *Sewells Automotive Marketing Review*, 6, 1–6.

Wells, P, P Nieuwenhuis & D G Rhys (1996b) Strategy in a saturated market. *Sewells Automotive Marketing Review*, 4, 1–5.

Wells, P, P Nieuwenhuis & D G Rhys (1996c) Scrappage incentives: green or greed? *Sewells Automotive Marketing Review*, 7, 3–5.

Wells, P, P Nieuwenhuis, D G Rhys & G Harbour (1997) *New Cars, Nearly New Cars and the Daily Rental Industry*. Cardiff: Centre for Automotive Industry Research.

Whitelegg, J (1993) *Transport for a Sustainable Future: The Case for Europe*. London: Belhaven.

Williams, H (1991) *Autogeddon*. London: Jonathan Cape.

Williams, K, C Haslam & J Williams, with A Adcroft & S Johal (1993) The myth of the line: Ford's production of the Model T at Highland Park, 1909–16. *Business History*, 35(3), 66–87.

Williams, K, C Haslam, S Johal & J Williams (1994) *Cars: Analysis, History, Cases*. Providence, RI: Berghahn Books.

Williams, P (1996) The current and future use of aluminium in car body structures. Paper presented at the 11th International Aluminium Conference, 8–10 September, Berlin.

Withers, J (1996) What's new? *The Independent Magazine*, 9 March, 24–8.

Womack, J & D Jones (1996) *Lean Thinking: Banish Waste and Create Wealth in Your Corporation.* New York: Simon & Schuster.

Womack, J, D Jones & D Roos (1990) *The Machine that Changed the World.* New York: Rawson Associates.

Worland, S (1994) A bit of class. *Mountain Biking UK*, January, 95.

Yanarella, E & R Levine (1992) Does sustainable development lead to sustainability? *Futures*, **24**(8), 759–74.

Zuckermann, W (1991) *End of the Road: The World Car Crisis and How We Can Solve It.* Cambridge: The Lutterworth Press.

Glossary of Terms and Abbreviations

2 + 2 A shorthand term for a vehicle body style that allows for two adults at the front and two 'occasional' seats at the back.

4 × 4 A car that is driven at all four wheels.

A pillar (also B, C and D pillars). A vehicle roof is separated from the body by up to four pillars each side, the A pillar being the one nearest the front of the vehicle. The A pillar defines the sides of the windscreen.

ACEA Association des Constructeurs Européens d'Automobiles, or the European Automobile Manufacturers Association: the automotive industry representative body for Europe, based in Brussels.

Aftermarket A collective term for the market for spare parts, servicing, maintenance and accessories for cars.

AGV Automatic guided vehicle.

AID Automotive Industry Data: a UK firm supplying data to the automotive industry.

AIV Aluminium-intensive vehicle: a vehicle designed to incorporate a larger amount of aluminium than is the norm. The Audi A8 is a recent example.

Bake hardening steel A form of cold-reduced sheet steel which is relatively easy to press, but which hardens in the elevated temperatures of the paint process. As a result, the steel is more resistant to dents under low-speed impacts.

BEV Battery electric vehicle: a vehicle that relies on traction batteries as its only source of motive power.

Big 3 A term commonly used to refer to America's major car-makers: General Motors, Ford and Chrysler.

Billion Throughout the book 'billion' is used to refer to 1000 million, i.e. US usage.

BIW Body-in-white: the fully welded monocoque car body before painting; usually defined as the structural part of the body without so-called hang-ons such as doors, bonnet/hood, bootlid/trunk or wings/fenders.

Blank A sheet of metal (usually steel) that has been pre-cut to the approximate shape required for pressing. All integrated transfer presses require pre-cut blanks.

Block exemption A provision by the European Commission whereby the practice of selective and exclusive distribution of cars via franchises is allowed to continue, even though it is in contravention of the Treaty of Rome.

BTU British thermal unit: a unit of energy.

CAIR Centre for Automotive Industry Research at Cardiff Business School, University of Wales, Cardiff, UK: the leading UK academic centre studying the economic and strategic aspects of the world motor industry.

CARB California Air Resources Board: the regulatory body which proposed a stepped programme to introduce zero-emissions vehicles onto the California market.

Carbon tax Commonly used term for any kind of tax aimed at the reduction of emissions of carbon dioxide through reduced consumption of fossil fuels.

Cd Coefficient of drag: one of the measures used in assessing the aerodynamic performance of shapes such as car bodies.

CFCs Chlorofluorocarbons: chemical compounds used in aerosol cans, refrigerators, air-conditioning systems and as cleaning agents in the electronics industry. They are directly implicated in the depletion of the stratospheric ozone layer and are being phased out under the provisions of the Montreal Protocol.

CNG Compressed natural gas: natural gas is compressed in order to reduce its volume so that it may be accommodated on board a vehicle in order to power its engine. Gradually becoming more popular as an automotive fuel, especially for urban use such as city buses; it produces less harmful emissions generally than petrol or diesel, although its energy density is lower, thus necessitating fairly bulky fuel tanks.

CRADA Collaborative Research and Development Agreements: a framework under which, in the US, companies and government research organisations may create consortia.

CVT Continuously variable transmission: a system which transfers power from the engine to the wheels without a fixed gearbox ratio.

Die A forming shape, usually constructed as pairs of mating dies and used to press steel, shape SMC, etc.

Downsizing A general move away from the larger, expensive segments of the car market towards small cars. Distinct from downsizing in the corporate sense, which usually entails a reduction in staff numbers.

Drivetrain An industry term sometimes used to describe the power transmission system from the clutch onward to the wheels, more widely used in commercial vehicles. See also *powertrain*.

EDI Electronic data interchange.

EFTA European Free Trade Area.

ELV End-of-life vehicle: a vehicle that has reached the end of its use-phase and is ready to be scrapped.

EOV Environmentally optimised vehicle.

ESRC Economic and Social Research Council: a UK government body which directs research through funding programmes.

EUCAR European Council for Automotive Research and Development: a forum for establishing and facilitating collaborative R&D in the European automotive sector.

EV Electric vehicle.

Evaporative emissions Emissions of volatile organic compounds as exemplified by petrol evaporating from a car's fuel system before combustion.

Extrusion A process whereby a material is forced under pressure through a shaping die to produce long sections. Widely used, for example, in aluminium glazing channels.

Finite-element analysis A computer-based means of predicting the structural performance of materials and components, and of the entire vehicle body. There are two common approaches: the beam model and the shell model.

Ford 2000 A major programme of internal restructuring within Ford whose major aim is the recapturing of economies of scale through world car platforms engineered within one of Ford's locations, but built worldwide in several locations.

FWD Front-wheel drive.

GDP Gross domestic product: a traditional economic and material measure of the total wealth generated by a nation.

GM General Motors.

GRP Glassfibre-reinforced plastic.

HPV Human-powered vehicle: anything that is powered by human effort but is not allowed to race under the rules of the Federation Internationale du Cyclisme.

HSS High-strength steel: a new family of steels showing greater strength than 'normal' cold-reduced steel.

Hybrid-powered vehicle A vehicle that has two distinct power plants, usually internal combustion and electric. The term is also sometimes used to refer to a body constructed from more than one material, such as steel and composites.

Hydroforming An industrial process where hollow metal sections are internally pressurised with fluid and then bent to shape. The internal pressure prevents the buckling that might otherwise occur. It is particularly suited to bending aluminium extrusions. A variant of this technique uses air instead of fluid; this technique was developed by ALCOA and is used for the Audi A8 semi-spaceframe.

ICDP International Car Distribution Programme: a programme funded by several car producers and a number of related interested bodies that is dedicated to the study of the future of car retailing and distribution; especially interested in the implementation of 'lean' practices in this sector.

Incrementalism A process of change in which there are cumulative marginal improvements leading to a step-by-step change towards an ultimate goal, rather than a revolutionary quantum leap.

IPCC Intergovernmental Panel on Climate Change.

IVHS Intelligent vehicle highway systems: the use of telematic systems to control and manage traffic, and aggregate and provide information to individual drivers.

JAMA Japan Automobile Manufacturers Association: westernised title for the Nihon Jidosha Kogyo Kai.

JD Power and Associates A US-based customer-tracking consultancy whose regular report on perceived quality of product and dealer bodies greatly influence the US market and the strategies of manufacturers selling in the US market. Has recently launched a tentative bridgehead in Europe in partnership with the BBC *Top Gear* programme and magazine in the UK.

KATECH Korea Automotive Technology Institute.

Kei cars The smallest segment in the Japanese car market, defined by external dimension and engine size (less than 660 cc).

Lean burn As opposed to 'normal' combustion which is stoichiometric (i.e. at 14.6 parts air to fuel), lean burn entails a greater proportion of air (i.e. usually over 20.0 parts air to fuel).

Lean production A system of changing the mass-production system such that all waste and non-value-adding activity is eliminated. It was inspired by the Toyota Production System and the term was popularised by the book *The Machine that Changed the World*.

LEV Low emissions vehicle, as defined by the California Air Resources Board.

LPG Liquid (or liquefied) petroleum gas: a by-product of the oil industry, normally fired off but increasingly used as an automotive fuel. Already widespread in some countries such as the Netherlands and South Korea. Also often used for forklift trucks, etc.

MCC   Micro Compact Car AG, the joint venture company between Mercedes and the SMH group owned by N. G. Hayek.

MEP   Member of the European Parliament.

MITI   Ministry of International Trade and Industry in Japan.

MMC   Mitsubishi Motor Company.

Model year   A term used in North America. The model year does not exactly correlate with the calendar year. Typically in the US market, plans and models for the forthcoming model year are established in September and October.

Monocoque   Also known as unibody, unitary body or integral body: a three-dimensional structure combining the roles of chassis and bodywork.

MPV   Multipurpose vehicle: term used in Europe for people-carriers, monovolumes, such as the Renault Espace; known in the US as minivans or garageable vans.

NASA   National Aeronautical and Space Administration, United States.

NiCad   Nickel–cadmium battery: a type of rechargeable battery increasingly used for smaller applications such as computer or mobile-phone batteries, but also under consideration for BEV use.

NVH   Noise, vibration and harshness: a collective term for unwanted intrusions into a vehicle cabin caused, for example, by vehicles resonating at certain frequencies.

OECD   Organisation for Economic Co-operation and Development: a club for rich countries which provides information through research studies, conferences, etc.

Paradigm   An overarching theory or body of practice and belief; within any one period most scientific endeavour will take place within the theoretical confines of the paradigm prevailing at that time, although over time, paradigms may 'shift', thus accommodating more radical change until a new paradigm becomes established.

Parallel hybrid   A vehicle with two power sources and which can be driven directly by either power source.

Parc   The number of vehicles in use or in circulation.

PCV   Positive crankcase ventilation.

Platform   Originally that part of the monocoque body that equates most with the chassis, i.e. the main structural part. This represents around 40% of the cost of the BIW and defines the main parameters such as suspension pick-up points, engine mountings, etc. Increasingly loosely defined.

PM 10   A group of particles found in automotive emissions, particularly from diesel engines, which are smaller than 10 μm and which are implicated in a number of respiratory illnesses.

PNGV   Partnership for a New Generation of Vehicles: a US collaborative programme involving public bodies and the Big 3 car-makers aimed at producing the car of the future. It was set up by the first Clinton administration and was inspired by the work of Lovins and the Rocky Mountain Institute.

Powertrain   A collective term for the engine, gearbox and transmission system of a vehicle.

PSA   Peugeot Société Anonyme: makers of Peugeot, Citroën and formerly Talbot cars.

Quarter panel   The section of the body above the rear wing and behind the rear door.

Regenerative braking   A braking system which transfers the energy required to stop a vehicle into stored battery energy.

RIM   Resin injection moulding: a system for moulding composite structures whereby the resin is mixed with smaller strands of reinforcing fibre before being injected into the mould; it combines the advantages of fibre-reinforced materials with those of injection moulding.

rpm   Revolutions per minute: a measure of engine speed.

RRIM   As for RIM, but with reinforcements.

RTM   Resin transfer moulding: a system for the moulding of composite panels whereby the fibre-reinforcement material is fitted in a mould, and then resin is poured or pumped into it. The Lotus VARI (vacuum-assisted resin injection) system is a variation on this which involves creating a vacuum inside the mould which 'sucks' in the resin, leading to better filling of the mould.

Sandwich steels   or laminate steels. A material made by sealing a thin layer of resin between two layers of sheet steel.

Scrappage incentives   Schemes introduced by governments whereby owners of old cars are offered a cash incentive to scrap that vehicle as long as they purchase a new one.

Seam welding   A process whereby two sheets of metal are joined along a continuous seam. Commonly achieved using electric resistance welding wheels, this area has received renewed interest following the development of laser seam welding.

Segment   A part of the total market at which a particular product is aimed; in the car market it is largely arranged by price and size. Segments are somewhat arbitrary and change over time, and may often be specific to an individual national market.

Series hybrid   A vehicle with two power sources, where only one source actually drives the vehicle directly. Typically, there will be a thermal engine (i.e. internal combustion) linked in series to an electric motor, which in turn provides the motive power to the wheels.

Spaceframe (chassis)   A three-dimensional structural chassis consisting of an arrangement of tubes creating a kind of cage to which the non-structural body panels are fitted.

Spot welding   A process whereby two or more metal sheets are joined at one point by resistance welds. The average contemporary car body is joined by a few thousand spot welds.

Springback   A phenomenon whereby pressed metal shows a tendency to return to its original shape; particularly a problem with recently developed 'high-strength' steels.

Strut brace   A tube across the engine bay of a car which connects the suspension turrets, thereby giving added torsional stiffness. Commonly used in rally cars. Sometimes referred to as engine brace.

SMC   Sheet moulding compound (or composite): a form of fibre-reinforced resin material which is pre-produced as a resin-impregnated rolled material. It can be unrolled and moulded by means of a pressing process, almost like sheet metal, at the start of the production process, thus reducing curing time. Increasingly used in truck cabs and car body panels.

SMMT   Society of Motor Manufacturers and Traders: the industry representative body in the UK.

SNCF   Société Nationale des Chemins de fer Français: the French national railway company.

Socio-technical paradigm   A combination of technological capabilities and social norms established over a long period of time.

SRIM   Structural reaction (or resin) injection moulding: a process which may be used in the construction of plastic panels.

Swing axles   A system of using one half-axle for each wheel whereby the centre is fixed, while the outer wheel-end can move up and down. Whilst providing good comfortable ride qualities, the system can cause a loss of grip and was implicated for this reason in Ralph Nader's damning book *Unsafe at Any Speed*.

Tailor-welded blanks   Flat sheets of steel made to approximately the shape required for pressing by joining together two or more separate sheets. The separate sheets may be of different thickness, have different coatings, etc. Increasingly practised using laser seam welding.

TLEV Transitional low emissions vehicle, as defined by the California Air Resources Board.

Torsional stiffness A means of measuring the 'efficiency' of a vehicle body in terms of the amount of effort required to twist the body out of shape.

TPS Toyota Production System.

ULEV Ultra low emissions vehicle, as defined by the California Air Resources Board.

US87 A series of emissions standards introduced in the US in 1987.

USCAR United States Council on Automotive Research: a collaborative forum for Chrysler, Ford and GM to co-ordinate R&D proposals.

VAG Volkswagen Aktiengesellschaft; makers of Volkswagen, Audi, Seat and Skoda cars, and commercial vehicles.

VOCs Volatile organic compounds: often harmful chemicals that can cause emissions through evaporation, such as the emissions of solvents from automotive paint plants.

VW Volkswagen.

Weld bonding A process for fixing materials together using both welding and some form of bonding involving adhesive; increasingly used to make structures involving aluminium and composites.

ZEV Zero-emissions vehicles, particularly as defined by Californian emissions legislation.

Index